# Synthesis Lectures on Engineering, Science, and Technology

The focus of this series is general topics, and applications about, and for, engineers and scientists on a wide array of applications, methods and advances. Most titles cover subjects such as professional development, education, and study skills, as well as basic introductory undergraduate material and other topics appropriate for a broader and less technical audience.

Praveen Cheekatamarla · Kyle Gluesenkamp ·
Stephen Kowalski · Zhenning Li · Saad Jajja

# The Role of Fuels in Transforming Energy End-Use in Buildings and Industrial Processes

Praveen Cheekatamarla
Energy and Science Technology Directorate (ESTD)
Oak Ridge National Laboratory
Oak Ridge, TN, USA

Kyle Gluesenkamp
Energy and Science Technology Directorate (ESTD)
Oak Ridge National Laboratory
Oak Ridge, TN, USA

Stephen Kowalski
Energy and Science Technology Directorate (ESTD)
Oak Ridge National Laboratory
Oak Ridge, TN, USA

Zhenning Li
Energy and Science Technology Directorate (ESTD)
Oak Ridge National Laboratory
Oak Ridge, TN, USA

Saad Jajja
National University of Sciences and Technology
Islamabad, Pakistan

ISSN 2690-0300  ISSN 2690-0327 (electronic)
Synthesis Lectures on Engineering, Science, and Technology
ISBN 978-3-031-45364-9  ISBN 978-3-031-45365-6 (eBook)
https://doi.org/10.1007/978-3-031-45365-6

© The Editor(s) (if applicable) and The Author(s), under exclusive license to Springer Nature Switzerland AG 2024

This work is subject to copyright. All rights are solely and exclusively licensed by the Publisher, whether the whole or part of the material is concerned, specifically the rights of translation, reprinting, reuse of illustrations, recitation, broadcasting, reproduction on microfilms or in any other physical way, and transmission or information storage and retrieval, electronic adaptation, computer software, or by similar or dissimilar methodology now known or hereafter developed.

The use of general descriptive names, registered names, trademarks, service marks, etc. in this publication does not imply, even in the absence of a specific statement, that such names are exempt from the relevant protective laws and regulations and therefore free for general use.

The publisher, the authors and the editors are safe to assume that the advice and information in this book are believed to be true and accurate at the date of publication. Neither the publisher nor the authors or the editors give a warranty, expressed or implied, with respect to the material contained herein or for any errors or omissions that may have been made. The publisher remains neutral with regard to jurisdictional claims in published maps and institutional affiliations.

This Springer imprint is published by the registered company Springer Nature Switzerland AG
The registered company address is: Gewerbestrasse 11, 6330 Cham, Switzerland

Paper in this product is recyclable.

# Preface

The author team at Oak Ridge National Laboratory has been working on technologies relevant to building energy transformation. As part of this research interest, we have been looking into the role of different primary energy sources for a sustainable energy transition. One of the questions of interest was to find out if the traditional and future fuels have any role in the decarbonization of all energy consumption sectors via electrification. In other words, under what scenarios do fuel-driven technologies offer benefits to the end user and simultaneously have comparable environmental impact? Specifically, what are the threshold electrical grid carbon intensities needed for fuel-based systems to be carbon competitive? If such scenarios exist, what is the relative performance of different fuel choices relevant for this energy transition and how the carbon intensity of these primary energy sources can be utilized in the decision-making process for immediate and long-term adoption. Given global society's goal of decarbonization, what is the appropriate choice from the various technologies that are available for fulfilling the heating and cooling needs of buildings? We hope this book provides the needed perspective for the readers in understanding the overall value proposition of different primary energy sources and thermal technologies.

| | |
|---|---|
| Oak Ridge, USA | Praveen Cheekatamarla |
| Oak Ridge, USA | Kyle Gluesenkamp |
| Oak Ridge, USA | Stephen Kowalski |
| Oak Ridge, USA | Zhenning Li |
| Islamabad, Pakistan | Saad Jajja |

# Introduction

This book examines what role fuels have in the transformation of end-use in buildings and industrial processes. Energy-efficient technologies are necessary to lower the carbon footprint for a transition towards clean energy in a sustainable manner. Efficient utilization of primary energy resources including renewables to support the current and future energy needs while targeting grid resiliency, energy, and environmental security, at an affordable cost is of significant value. Analysis of configurations consisting of heat pumps, fuel-driven thermal providers, and power systems is presented. Sensitivity of electrical grid's carbon intensity towards carbon footprint in comparison with fuel-driven technologies is necessary to recognize the true value proposition of currently available energy solutions for different end consumers. Similarly, the role of low-carbon, zero-carbon, and carbon-negative fuels such as power to gas, power to liquid, hydrogen, biogas, etc., in conjunction with polygeneration technologies are discussed. Transformation of the primary energy resources from conventional fossil fuels to renewable fuels or electricity will have a significant impact on the overall carbon footprint of various end-use sectors, including buildings. Hence, this book also examines two different scenarios focused on sensitivity of the pace of decarbonization of electrical grid and fuel supply on operational energy-related carbon emissions.

# Contents

1   **Evolution of the Energy Landscape** .................................... 1

2   **Thermal Technologies** ................................................ 5
    2.1   Residential Applications .......................................... 5
        2.1.1   Residential Cooling ....................................... 5
        2.1.2   Residential Heating ....................................... 6
        2.1.3   Residential Thermal Energy Storage .......................... 8
        2.1.4   Residential Potable Water Heating .......................... 8
        2.1.5   Other Residential Applications ............................. 9
    2.2   Commercial Applications ......................................... 9

3   **Global Primary Energy Sources and Their Carbon Intensity** .............. 11
    3.1   Global Primary Energy Sources .................................... 11
    3.2   Carbon Intensity of Primary Energy Sources ......................... 13
    3.3   Grid Emission Factor ............................................ 15

4   **Calculation Approach and Assumptions** ............................... 19
    4.1   Heating Technologies ............................................ 21
        4.1.1   Conventional Electrical Resistance and Heat Pump Systems, Direct Fuel Fired and Thermally Driven Sorption Systems ....... 22
        4.1.2   Self-powered Devices ..................................... 22
        4.1.3   Dual-Fuel Devices ........................................ 22
        4.1.4   CHP Systems with and Without Thermal Energy Storage (TES) ................................... 23
        4.1.5   CHP Systems with Sorption Heat Pump ...................... 24
        4.1.6   Blackstart Heat Pump Systems with and Without Thermal Energy Storage .................................. 24
    4.2   Cooling Technologies ............................................ 25
        4.2.1   Conventional and Sorption Air Conditioners ................... 26
        4.2.2   CHP Configurations with Conventional AC or Sorption AC ...... 26

| 5 | **Effective Carbon Footprint of Heating Technologies** | 29 |
|---|---|---|
| | 5.1 Carbon Footprint of Non-power Generating Thermal Technologies | 29 |
| | 5.2 Carbon Footprint of Power Generating Thermal Technologies | 32 |
| 6 | **Effective Carbon Footprint of Cooling Technologies** | 35 |
| | 6.1 Carbon Footprint of Non-power Generating Cooling Technologies | 35 |
| | 6.2 Carbon Footprint of Power Generating Cooling Technologies | 36 |
| 7 | **Effective Carbon Footprint of Industrial Heating Technologies** | 39 |
| | 7.1 Low Temperature Processes | 40 |
| | 7.2 High Temperature Processes | 41 |
| 8 | **Decarbonization Scenarios** | 45 |
| | 8.1 Building Heating Technologies | 45 |
| | 8.2 Building Cooling Technologies | 47 |
| | 8.3 Industrial Heating Technologies | 48 |

**Concluding Remarks** ............................................................. 51

**Nomenclature** ...................................................................... 53

**References** .......................................................................... 57

# About the Authors

**Praveen Cheekatamarla** (Ph.D., Chemical Engineering) is a Senior Researcher at Oak Ridge National Laboratory (ORNL) focusing on developing energy-efficient building equipment and appliances. He has over 16 years of industrial experience in energy-efficient product development and served as a technical leader and principal engineer. He was the Director of Research and Product Development at Atrex Energy and was the principal developer of the fuel cell technologies, enabling commercial deployment of more than 800 fuel cell remote power generators in North America. His primary areas of expertise include power systems, surface science, material properties, thermo-chemical processes, energy modeling, systems analysis and integration, and energy-efficient process/product development. He has more than 70 publications and presentations and 14 invention records, and two patents to his name. He is an active peer reviewer, editorial board member, and guest editor for technical journals, and technical advisor for startups.

**Kyle Gluesenkamp** is a Senior R&D Scientist and serves as ORNL's Subprogram Manager for Thermal Energy Storage. He has 15 years of experience and deep expertise in sorption technologies, thermal storage and phase change materials, residential appliances (clothes dryers and dishwashers), combined heat and power, heat transfer, psychrometrics, and experimental prototype evaluation. He has published two book chapters, 45 journals, and over 90 conference articles and reports, and has four granted patents.

**Stephen Kowalski** is a Senior R&D Staff Member in the Multifunctional Equipment Integration Group at Oak Ridge National Laboratory. His work at the laboratory is centered on new technologies for residential, commercial, and industrial HVAC. These technologies have included novel methods of vapor compression for air conditioners and heat pumps as well as vapor compression cycle alternates, new compression technologies, and carbon capture. Prior to working at ORNL, he worked for over 25 years for Trane Technologies. There, he focused on product and technology development for residential and commercial gas furnaces and air conditioners. He has been an active member of ASHRAE participating in Technical and Standard Project Committees. He was also a member of the GAMA furnace engineering subcommittee and is a member of the technical subcommittee for CSA standards P.2 and P.8 and of the AHRI Standards Consensus Body focused on heating standards. He has also been an Adjunct Professor of engineering technology at Austin Peay State University in Clarksville, Tennessee.

## About the Authors

**Zhenning Li** is a Staff Scientist in Oak Ridge National Laboratory. He received his Ph.D. from the University of Maryland in 2019 and received his B.S. from Shanghai Jiao Tong University in 2014. He specializes in numerical modeling and design optimization of building equipment, especially heat pump technologies. He serves as the vice-chair of ASHRAE Technical Committee 1.13-Optimization.

**Saad Jajja** is a Postdoctoral research associate in the Building and Transportation Science Division at the Oak Ridge National Laboratory. He obtained his doctoral degree in Mechanical Engineering from Oregon State University with a focus on thermal-fluid sciences. His doctoral work was focused on experimental and analytical characterization of turbulent heat transfer of supercritical carbon dioxide in the proximity of the pseudo-critical point. At Oak Ridge, he works on the multi-phase heat and mass transport fundamentals of low global warming potential (GWP) refrigerants. His work has been published in several top thermal-fluid science journals which include the *International Journal of Heat and Mass Transfer*, *Experimental Thermal and Fluid Sciences*, and the *International Journal of Refrigeration*.

# List of Figures

| | | |
|---|---|---|
| Fig. 1.1 | Trends in the fuel source for electricity generation and the emissions associated with this sector | 2 |
| Fig. 1.2 | Maximum theoretical efficiencies of fuel cell-engine hybrid system | 3 |
| Fig. 3.1 | Primary energy consumption by source, 1965–2022 | 12 |
| Fig. 3.2 | Historical global primary energy consumption by source | 14 |
| Fig. 3.3 | $CO_2$ emissions for different fuels | 15 |
| Fig. 3.4 | Global carbon intensity of electricity in 2022 | 16 |
| Fig. 5.1 | Carbon footprint associated with thermal loads in a building consuming 100 kWh/day supplied with electrical grid or natural gas fuel | 31 |
| Fig. 5.2 | Carbon footprint associated with thermal loads in a building consuming 100 kWh/day supplied with electrical grid (700 $gCO_{2e}$/kWh) and fuel with different carbon intensities | 31 |
| Fig. 5.3 | Combined electrical and thermal load carbon footprint in a building consuming 100 kWh/day thermal energy and 30 kWh/day electrical energy supplied with electrical grid (700 $gCO_{2e}$/kWh) and fuel with different carbon intensities | 33 |
| Fig. 5.4 | Combined electrical and thermal load carbon footprint in a building consuming 100 kWh/day thermal energy and 30 kWh/day electrical energy supplied with electrical grid (50 $gCO_{2e}$/kWh) and fuel with different carbon intensities | 34 |
| Fig. 6.1 | Carbon footprint associated with cooling load in a building supplied with electrical grid at a carbon intensity of 700 $gCO_{2e}$/kWh and fuels with different carbon intensities | 36 |
| Fig. 6.2 | Carbon footprint associated with cooling load in a building supplied with electrical grid at different carbon intensities, compared with a sorption system fueled by natural gas at a carbon intensity of 180 $gCO_{2e}$/kWh | 37 |

| Fig. 6.3 | Combined electrical and cooling thermal load carbon footprint in a building supplied with electrical grid at different carbon intensities and onsite power generation systems fueled by natural gas at a carbon intensity of 180 $gCO_{2e}$/kWh ......................... | 38 |
|---|---|---|
| Fig. 6.4 | Combined electrical and cooling thermal load carbon footprint in a building supplied with electrical grid at a carbon intensity of 700 $gCO_{2e}$/kWh and fuels with different carbon intensities ......... | 38 |
| Fig. 7.1 | Carbon intensity of thermal load provided by non-power generating heating technologies supplied with different electric grid carbon intensity factors. Natural gas as the fuel based primary energy ........ | 40 |
| Fig. 7.2 | Carbon footprint of thermal load provided by non-power generating heating technologies supplied with different carbon intensity factors of the fuel supplied and electrical grid with carbon intensity of **a** 700 $gCO_{2e}$/kWh (carbon heavy), **b** 50 $gCO_{2e}$/kWh (clean/renewable) ...... | 42 |
| Fig. 7.3 | Industrial heating equipment for processing metals, fabric, plastic, glass, ceramic, food, wood, coatings etc. (with permission from Trimac Industrial Systems) ................................. | 43 |
| Fig. 7.4 | Carbon footprint of combined thermal and electrical in an industry supplied with different carbon intensity factors of the fuel along with electrical grid at carbon intensity of: **a** 700 $gCO_{2e}$/kWh (carbon heavy), **b** 400 $gCO_{2e}$/kWh (moderately polluting), **c** 50 $gCO_{2e}$/kWh (clean/renewable) ................................... | 44 |

# List of Tables

| | | |
|---|---|---|
| Table 3.1 | Largest primary energy producing countries in 2022 | 13 |
| Table 3.2 | Grid emission factor for major primary source to generate electricity [13] | 16 |
| Table 4.1 | Assumptions and operational conditions assumed in the energy and carbon footprint analysis presented | 20 |
| Table 5.1 | Carbon intensity of thermal load provided by non-power generating heating technologies supplied with different electric grid carbon intensity factors. Natural gas as the fuel based primary energy | 30 |
| Table 8.1 | Electric grid decarbonization scenario—combined electrical and thermal energy carbon footprint in a building consuming 100 kWh/day thermal energy and 30 kWh/day electrical energy. Fuel carbon intensity of 180 g$CO_{2e}$/kWh | 46 |
| Table 8.2 | Fuel grid decarbonization scenario—combined electrical and thermal energy carbon footprint in a building consuming 100 kWh/day thermal energy and 30 kWh/day electrical energy. Electric grid carbon intensity of 400 g$CO_{2e}$/kWh | 46 |
| Table 8.3 | Electric grid decarbonization scenario—thermal energy carbon footprint of dual fuel system operated at different primary energy operational ratios. Fuel carbon intensity of 180 g$CO_{2e}$/kWh | 47 |
| Table 8.4 | Fuel grid decarbonization scenario—thermal energy carbon footprint of dual fuel system operated at different primary energy operational ratios. Electric grid carbon intensity of 400 g$CO_{2e}$/kWh | 47 |
| Table 8.5 | Electric grid decarbonization scenario—combined electrical and cooling energy carbon footprint. Fuel carbon intensity of 180 g$CO_{2e}$/kWh | 48 |
| Table 8.6 | Fuel grid decarbonization scenario—combined electrical and cooling energy carbon footprint. Electric grid carbon intensity of 400 g$CO_{2e}$/kWh | 48 |

| | | |
|---|---|---|
| Table 8.7 | Electric grid decarbonization scenario—combined electrical and thermal energy carbon footprint. Fuel carbon intensity of 180 $gCO_{2e}$/kWh | 49 |
| Table 8.8 | Fuel grid decarbonization scenario—combined electrical and thermal energy carbon footprint. Electric grid carbon intensity of 400 $gCO_{2e}$/kWh | 49 |

# Evolution of the Energy Landscape

The story of human civilization is one of constant progress that has seen the descendants of wandering hunter-gatherers land on the surface of the moon. The ancients utilized the energy stored in their muscles to launch projectiles to hunt whereas, the moderns created engines that unlock the energy in chemical bonds to allow them to take trips to the heavens. Our forefathers have always known subconsciously that the source of all energy for our planet is the sun. They recognized the importance of sun in human affairs and worshiped it as a god—Ra was the Egyptian god of the sun. Besides worship, they also used it as a weapon of war. Legend has it that Archimedes used mirrors to concentrate sunlight to set ablaze Roman warships.

The great act by Prometheus in stealing the fire from the gods and gifting it to humanity, gave our ancestors a tool to start unlocking sun's energy contained in the chemical bonds of organic matter. However, it would take many generations of Prometheus's descendants to harness this fire which would enable Neil Armstrong and Buzz Aldrin to walk on the surface of the moon in the July of 1969. Over the passage of time our ancestors have exploited sun's energy from diffuse sources to very concentrated sources. An example of using a diffuse source would be feeding draft animals a plant-based diet and then relying upon these animals to do manual labor. These animals can typically provide a steady state power of 5 kW [2]. Invention of steam engines allowed us to extract the sun's energy from more concentrated sources, such as coal, which then started the industrial revolution.

Once we learned how to do that, the story of progress is mainly improving these devices and inventing their derivatives. Steam driven engines have greatly improved over the years and are widely used in power generation industry in the Rankine cycle configuration. Additionally, the invention of gas-turbine engines has enabled humanity to extract

**Fig. 1.1** Trends in the fuel source for electricity generation and the emissions associated with this sector

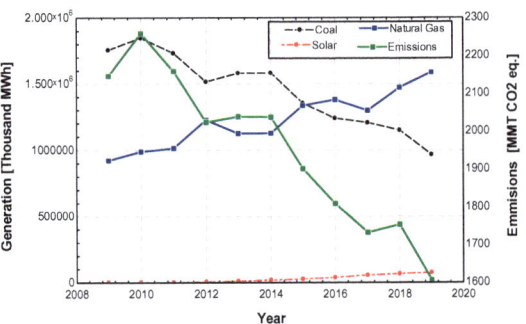

and control vast amounts of power—A pilot of a Boeing's 747-400 controls 45 MW of power at his fingertips [2]. The drive towards improvement in these energy conversion devices is and was mainly driven by economic reasons but in recent years, concerns about the effect of burning fossil fuels on the environment are also being considered during the design process. Improvements in efficiency, mainly driven by economic reasons, have environmental benefits as a by-product.

In the last decade (2009–2019), the US power generation market has transitioned to natural gas as a dominant fuel source while at the same time, there has been a reduction in the emissions associated with electricity generation sector. These trends are shown in Fig. 1.1 which was plotted by the data obtained from [5, 6]. This transition was mainly supported by two factors. Firstly, increase in the shale gas production from 3110 billion cubic ft. in 2009 to 25,556 billion cubic ft. in 2019 [4] resulted in a price reduction for natural gas as a fuel source. Secondly, higher efficiencies associated with natural gas-fired power plants result in increased utilization of these assets by plant operators to save costs. In 2019, the heat rate (BTU used to produce 1 kWh of electricity) for combined cycle natural gas plants was 7633 BTU/kWh which translates to an efficiency of 44.7% [3].

Further improvements in generation efficiency, coupled with cheap natural gas, can serve as an economic incentive to increase usage of natural gas as a fuel source. Based on the recent trends of the last decade, this might prove to be a more pragmatic approach to reduce the emissions associated with the electricity generation sector. General Electric's latest 9HA.02 gas turbine is reported to have efficiency of 64.3% (LHV) in a combined cycle arrangement ($2 \times 1$) [1, 7]. These 'H' class turbines have firing temperatures (Peak working temperature in Brayton cycle) higher than 1430 °C and will be used in large, centralized power stations (Output in the order of MW).

Continued improvements in power generation technologies, typically large, centralized designs, may be negated by the losses encountered elsewhere in the energy usage chain. For instance, the losses incurred in the electricity transmission grid are about 5% from 2015 through 2019 in the US. Additionally, these centralized plants might have poor resiliency under certain conditions—dramatic weather or terrorist events that can leave big parts of the country without power. Decentralized technologies can help to counter

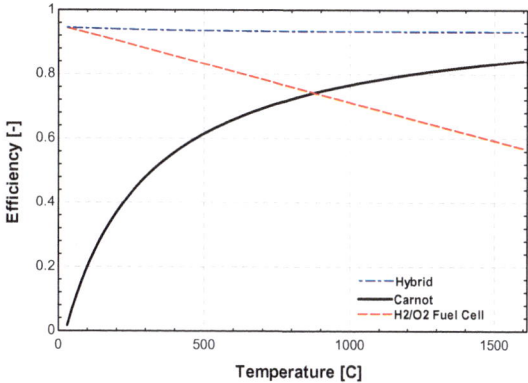

**Fig. 1.2** Maximum theoretical efficiencies of fuel cell-engine hybrid system

these issues and should be considered as a serious contender in our future energy mix. In many of these decentralized systems, there is also potential to achieve efficiencies associated with large, centralized power generation systems.

One approach that can avoid these losses while maintaining high generation efficiency is to couple a Solid Oxide Fuel Cell (SOFC) with a gas turbine in small scale (~100 kW) distributed systems. It is not clear if these systems will ever replace the existing combined cycle centralized power plants, however, they do present an option that can be pursued to diversify the generation mix. From a thermodynamic perspective, these hybrid systems can achieve higher efficiencies, as shown in Fig. 1.2. Some of these units have been built, for instance, a 220 kW hybrid system, designed and built by Siemens-Westinghouse, demonstrated an efficiency (net AC/LHV) of 53% [8].

Finally, the last link in the energy usage chain is the end-use appliances that we use in our houses and for transportation purposes. Efficiency gains in these devices by improving component design and integration of smart controls will help us realize the economic and environmental benefits, to the full extent, of the highly efficient prime movers that humanity has developed.

In this brief discussion, we have not talked about integrating renewable power sources into the current energy mix. This is not by accident since it must be stressed that the major gains achieved by our civilization have been focused on building and improving engines to extract and convert the energy contained in fossil fuels to either electricity, propulsion, or mechanical shaft work for ground transportation. These gains in engine performance which lead to a lower cost associated with energy conversion processes are largely responsible for improving the standard of living of people. Access to cheap energy and reliable engines allow people the freedom of movement to conduct economic activities and run business enterprises without energy bills eating the income flow.

This economic growth, strongly coupled with cheap and reliable energy conversion systems, creates disposable personal and national income that can then be routed to adoption and development of new technologies that may replace the fossil fuel powered prime movers in the future. For instance, a certain amount of disposable income allows people to buy electric vehicles which may lead to this industry benefiting from economies of scale, leading to lower costs and ultimately more adoption. Similarly, the nation, as a whole, should have enough income to spare to spend upon highly risky research and development of the next generation of prime movers.

At this point, only the developed nations have enough wealth to pursue these alternative technologies. The underdeveloped parts of the world still need to rely on cheap and abundant fossil fuel powered engines to power their economies and once they have enough national and personal income only then they can effectively adopt the new technologies being developed by wealthy nations. This aspect is one of the most significant challenges where sustainability and environmental protection are at odds with each other. Decision makers should be cognizant of the economic realities of the developing world when selecting the technologies to power their economies. Renewable power can supplement traditional base load fossil fuel power systems but, any additional costs that it will impose on the consumers should be carefully considered. Adoption of new technologies should happen in an organic fashion, with raising of living standards and strong economies as priorities while being environmentally responsible within the constraints of country's economic boundaries. We hope that the technologies and perspectives discussed in this book can help global readers and policy makers adopt suitable technologies and practices that are most suited to their particular region and economic situation.

# Thermal Technologies 2

## 2.1 Residential Applications

Several different types of thermal technologies are used for heating and cooling, water heating, and other miscellaneous applications in North American homes. Often the type of technology that is chosen is based on the type of construction and the geographical region of the continent. For air conditioning, the most common types of products are forced air, though chilled water based systems are also possible. There are more choices for heating, however. Forced air heating through ducts is the most common method of delivering residential heat, but other types of heating would include hydronic systems that use hot water or steam that is distributed from a boiler to radiators or baseboard convectors, devices installed within a room, through a wall or in a wall, baseboard electric resistance convectors, forced-air or natural convection wood-, coal-, or pellet-fueled stoves.

### 2.1.1 Residential Cooling

Common methods of cooling provided by forced air require a central air conditioner. This could take the form of a split system with the evaporator coil co-located with the central furnace such that the furnace fan is used to pull the return air from the conditioned space to the furnace, then push the return across the evaporator coil to be cooled and dehumidified, then returned to the conditioned space by a system of ductwork. Cooling of the evaporator coil can be achieved either by vapor compression or a sorption system. In a typical residential split system, the compressor and condenser coil are in a separate piece of hardware that sits outside the conditioned space to communicate and transfer heat to the ambient air, while the expansion device for the cooling application, most often a thermal expansion valve (TXV) but occasionally an electronic expansion valve (EEV)

© The Author(s), under exclusive license to Springer Nature Switzerland AG 2024
P. Cheekatamarla et al., *The Role of Fuels in Transforming Energy End-Use in Buildings and Industrial Processes*, Synthesis Lectures on Engineering, Science, and Technology, https://doi.org/10.1007/978-3-031-45365-6_2

is located with the evaporator coil. In addition to a split system, all the components can be located outside the structure in a single-packed product equipped with ducts to transfer the return air from the structure to the unit, then the supply air from the unit back to the structure. These systems are most often fully electric, with electricity from the grid being used to power the compressor, indoor and outdoor fans, and the controls. However, systems have been marketed that use a small internal combustion engine driving a semi-hermetic compressor for use in a vapor compression system. Therefore, a vapor compression system could be fossil-fueled.

Less commonly, cooling of the evaporator coil can be achieved with an absorption system that uses either a fossil-fueled burner or electric heat system. Though these systems have been marketed for many years, for various reasons they have not yet developed a large market. However, among their advantages are that they can use combustion of a fossil fuel to create a cooling effect.

Another method of residential cooling is using an evaporative cooler, sometimes called a "swamp cooler". This product works well in hot dry climates, such as the western United States. It uses the evaporation of liquid water into a vapor. This process uses the thermal energy in the air stream, resulting in lower temperature air that can be supplied to the residence.

### 2.1.2 Residential Heating

Residential heating can use either combustion of a fuel or electricity from the grid. A common method for heating is the use of forced air. This simplifies having both heating and cooling in a residence because the same set of duct work can be used for either application in a straightforward manner. The heat to be distributed through the ductwork is generated in the furnace either by combustion of natural gas, propane, fuel oil, coal, or wood or from electrical resistance elements. In the combustion applications, the combustion products are separated from the circulating air by a heat exchanger, while a separate heat exchanger is not required for heat generated by electric resistance elements. For combustion furnaces, though the source of heat is the combustion of a fuel, the furnace does have a requirement for electric power to drive the circulating fan, combustion fan (if equipped), and controls. However, the amount of power in the form of electricity required for these products is a small fraction of the amount of power consumed in the form of the combustion of the fuel. A self-powered furnace could then be developed to scavenge this small power fraction from the combustion of the fuel to be converted by some means into the electricity necessary for the fans and controls. If even more electricity can be generated, it can be used to power a vapor compression cycle, resulting in a self-powered air conditioner to go along with the self-powered furnace. These systems could be combined with some type of battery such that they would be able to generate electricity not only for operation, but excess electricity that could be stored to power controls during the off-cycle and even to power the fans necessary to start the combustion process in the next cycle. In this way, the product would not require connection to the electricity grid at all. A product with the ability to operate separate from the grid and store sufficient power to restart itself can be referred to as a black start appliance. Electrical storage could be combined with various appliances to create black start heat pumps or black start air conditioners. Examples of methods for the conversion of thermal energy to electrical energy could be thermophotovoltaic cells or thermo-electric generation.

In addition to heat being provided by the combustion of fuels through ductwork by a furnace, heat can be supplied by non-ducted or in-space appliances. Examples include room heaters and wall or floor furnaces. Like central furnaces, heat for these appliances is generated by either combustion of a fuel or electric resistance elements. Also similar, they require electricity from the grid to power fans and controls. Finally, in-space appliances with the combustion of fuels as their heat source could be made self-powered through some method to convert the heat of combustion to electricity to power fans and controls using methods as described above.

Another method to provide heating through ductwork or in an in-space application is using either the vapor compression cycle or through a sorption process as a heat pump. If heat is delivered by ductwork, the appliance can be either in a split construction or a single packaged product as discussed above. In the former case, the airflow for the indoor section is provided by an air handler or fan-coil that contains the indoor fan and indoor coil along with the expansion device needed for cooling operation. The indoor coil serves as the condenser for heating operation but can also be the evaporator coil for coiling operation by reversing the flow of refrigerant through the refrigerant tubing. The expansion device that is necessary for heating operation is in the outdoor unit. As with the cooling operation, in a single package product, all components are in the unit sitting outside the residence, again connected to the conditioned space by the return and supply ducts. The vapor compression cycle can use electricity from the grid to power a hermetic compressor or it can use a fuel to power an engine which in turn powers a semi-hermetic compressor. Heat pumps using electricity are attractive from an efficiency viewpoint as they have a coefficient of performance greater than one, which is not possible with a fuel-fired appliance. Heat pumps can use the ambient air as a heat source, which is referred to as an air-source heat pump. Alternately, heat pumps can use the ground as a heat source, which is referred to as a ground-source heat pump, or they can use a body of water as the heat source, which is referred to as a water source heat pump.

Commonly, heat pumps suffer from reduced heating capacity as the temperature of the heat source is reduced. As increased efficiency would be desired with decreasing ambient temperature, vapor compression heat pumps have a need for supplementary heat capability. This is commonly provided in the form of electric resistance heat elements. Another form of supplementary heat is provided by a combustion-fueled furnace. This can sometimes be called a dual-fuel system as electricity can be considered the "fuel" for the heat pump and a natural gas, propane, or fuel oil would be the fuel for the furnace. Additionally, this type of system can be referred to as a hybrid system. Currently, control systems for hybrid heat pumps do not allow the simultaneous operation of vapor compression heat pumping and combustion heating. However, if the technical challenges can be overcome, there are advantages to seamlessly changing between heat pump and combustion heating. For example, the fraction of vapor compression heat pump to combustion heat can be varied to use more combustion heat if electricity prices rise in times of high usage or if the power plant supplying electricity to the grid must rely on higher carbon-producing forms of electricity generation, such as coal combustion. In the latter case, signals from the electric utility can be received by the appliance to allow the operation that results in the lowest total carbon emissions.

Though this type of product is not currently commercialized, a "black-start" self-powered heat pump could be a method of residential (or commercial) space heating. The term "black-start" indicates that the product does not need any external power source to start. A self-powered product, such as mentioned previously, could be configured with a battery to store power during its operating cycle that can then be used to start the unit during its next cycle. This type of product could easily be integrated with simultaneous operation of vapor compression and combustion heating.

Another product that is not yet widely available for residential use would combine the function of heating with producing power for the whole home. This type of product, referred to as combined heat and power (CHP), is used for larger applications, typically in various industrial sectors. However, with the appropriate economics of initial purchase and operating costs, it could be applied residentially.

### 2.1.3 Residential Thermal Energy Storage

With the potential to move more heating capacity from fossil-fueled appliances to heat pumps, which is a desired goal in the decarbonization of energy use, thermal energy storage (TES) may become more important. The TES could be charged during times of low electricity grid usage, then discharged during times of high electricity grid usage. In this way, the peak of electricity usage could be reduced, eliminating the need for more electricity generation. This is an emerging area of energy conservation research. TES could be combined with any of the other technologies other than vapor compression heating. A combination of TES with a CHP system is one example.

### 2.1.4 Residential Potable Water Heating

Thermal technologies are also used in various ways to heat potable water for residential applications. Fuel-fired water heaters use the heat derived from the combustion of natural gas, propane, or fuel oil. These water heaters can be storage or tank water heaters that heat and store a large quantity of water on the order of 10s of gallons at the desired use temperature. Storage water heaters can also be powered by electricity from the grid, either by electric resistance heating elements or by a combination of a vapor compression cycle with electric resistance heating elements also installed for recovery at high usage rates. In addition to storage type water heaters, thermal technologies are also used for on-demand waters that do not store the heated water, but instead are equipped controls to turn on the burners or electric heating element at the time of demand, then off until needed again. Necessarily the maximum power input to these products is significantly higher than that of storage water heaters. Boilers used for residential heating can also be equipped with controls such that they can heat potable water for either on-demand or storage applications. Though it is less common, other thermal technologies such as absorption systems, could also be used for heating potable water in residential applications.

Heating potable water presents interesting opportunities for thermal energy storage as the tank of heated water itself can be the storage system. As the tank storing the heated water is well-insulated, the controls of the water heater can be configured to receive signals from the electric grid to heat the water earlier than needed such that it can be turned off when the electric grid usage is unusually high.

### 2.1.5 Other Residential Applications

Though they are not specifically part of this analysis, there are other uses of thermal technology in residential applications. These would include kitchen ranges and ovens, clothes washers and dryers, pool heaters, driveway snow melt appliances, etc. All of these can use a combination of fuel-fired combustion, electric resistance heat, vapor compression, and sorption technologies, among others. Additionally, thermal energy storage can be applied to these applications.

## 2.2 Commercial Applications

Thermal technologies are also used for commercial applications, specifically for space conditioning and potable water heating. The technologies used for smaller commercial applications like strip malls, convenience stores, or restaurants are very similar to the technologies used for residential applications. Single package products are very common. These feature vapor compression components for cooling and heating, but can also use combustion of fuels, such as natural gas, propane, or fuel oil or electric resistance elements for heating. Split systems can also be used for commercial applications, but these products do not typically use the combustion of a fuel for heating. There are many methods for space conditioning of larger commercial applications, but one example is they can also use single package products. These applications can also include the use of variable air volume (VAV) boxes located in the ceiling cavity above conditioned rooms that are equipped with dampers, fans, and heating elements to allow more precise control of temperature in zoned systems. More common in larger commercial applications, but more commonly use engineered systems made of combinations of a chiller and one or more catalog or engineered air handlers.

# Global Primary Energy Sources and Their Carbon Intensity

## 3.1 Global Primary Energy Sources

In energy statistics primary energy (PE) refers to the first stage where energy enters the supply chain before any further conversion or transformation process. Primary energy consists of:

- fossil, using coal, crude oil, and natural gas;
- nuclear, using uranium;
- renewable, using biomass, geothermal, hydropower, solar, tidal, wave, wind, and among others.

Figure 3.1 shows the global primary energy consumption by source. In 2022, the global primary energy distribution was as follows [9]: coal 27.6%; oil 31.6%; gas 25%; nuclear 4.4%; hydropower, 7%; wind 2.6%; solar 1.4% and other renewables 0.5%. Approximately 16% of our energy was derived from low-carbon sources, with nuclear power contributing slightly over 4% and the remaining 12% generated from renewable technologies. Table 3.1 shows the largest primary energy producing countries in 2022 [11].

Figure 3.2 resents a historical overview of global primary energy consumption, tracing back to 1800 AD. In the pre-industrial age, the primary source of energy predominantly consisted of traditional biomass, including the combustion of solid fuels such as timber, agricultural residue, or charcoal. However, the advent of the Industrial Revolution witnessed a significant upsurge in the utilization of coal, which accounted for approximately half of the world's energy requirements by the dawn of the twentieth century, with the remaining half still reliant on biomass. As the twentieth century unfolded, the global energy portfolio became increasingly diversified, introducing oil, gas, and subsequently,

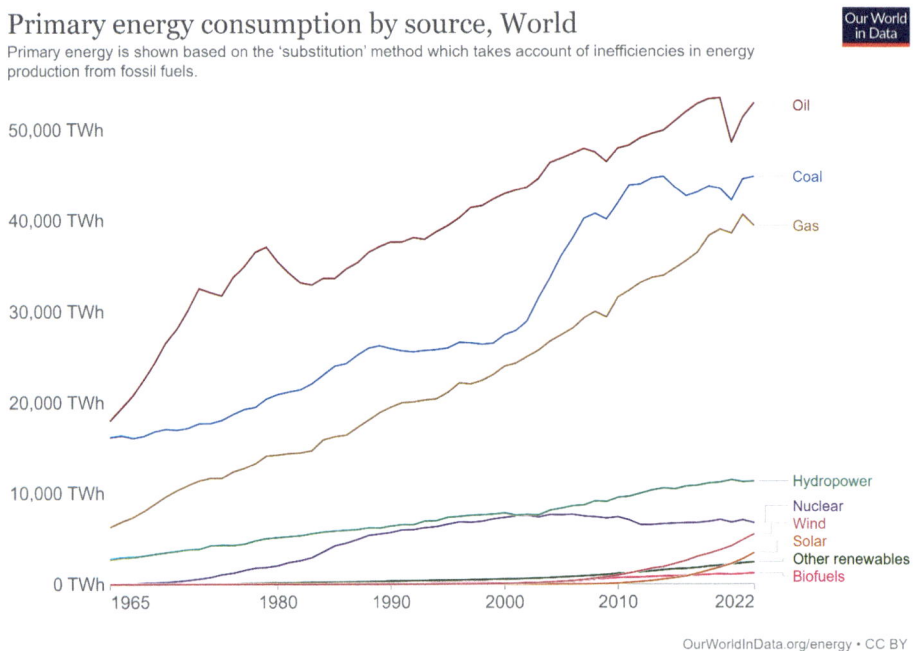

**Fig. 3.1** Primary energy consumption by source, 1965–2022

hydropower into the energy matrix. Nuclear energy, a relatively contemporary source, was not incorporated until the 1960s. The advent of what is often labeled as 'modern renewables,' namely solar and wind energy, did not occur until the latter part of the twentieth century, specifically the 1980s.

An important observation from this 200-year trajectory of global energy consumption is the incremental pace at which energy transitions have historically occurred. It has often taken several decades, if not a century, for a given energy source to gain preponderance. While this has been the norm historically, recent indicators suggest an accelerating trend in energy transitions. For instance, in the UK, coal accounted for nearly two-thirds of electricity generation in 1990, which dwindled to just under one-third by 2010. In the subsequent decade, this figure plummeted to approximately 1%. While past energy transitions have been marked by a slow pace, it is not an inevitable characteristic for future transitions.

As the long-term change for fossil fuel consumption, the IEA Sustainable Development Scenario [11] projects a significant reduction in fossil fuel consumption by 2040, driven by a shift to renewables, energy efficiency improvements, and electrification of various sectors. However, this transition is contingent on policy measures, technological advancements, and societal changes towards sustainable practices.

## 3.2 Carbon Intensity of Primary Energy Sources

**Table 3.1** Largest primary energy producing countries in 2022

|  | Total | Coal | Oil and gas | Renewable | Nuclear |
|---|---|---|---|---|---|
| China | 2,950 | 71% | 13% | 10% | 6% |
| United States | 2,210 | 13% | 69% | 8% | 10% |
| Russia | 1,516 | 16% | 78% | 2% | 4% |
| Saudi Arabia | 610 | 0 | 100% | 0 | 0 |
| Iran | 354 | 0 | 99% | 0 | 1% |
| United Arab Emirates | 218 | 0 | 99% | 0 | 1% |
| India | 615 | 50% | 11% | 33% | 6% |
| Canada | 536 | 5% | 81% | 10% | 4% |
| Indonesia | 451 | 69% | 17% | 14% | 0 |
| Australia | 423 | 64% | 33% | 3% | 0 |
| Brazil | 325 | 1% | 55% | 42% | 2% |
| Nigeria | 249 | 0 | 47% | 53% | 0 |
| Algeria | 150 | 0 | 100% | 0 | 0 |
| South Africa | 151 | 91% | 1% | 8% | 0 |
| Norway | 214 | 0 | 93% | 7% | 0 |
| France | 128 | 0 | 1% | 34% | 65% |
| Germany | 102 | 27% | 3% | 47% | 23% |

## 3.2 Carbon Intensity of Primary Energy Sources

Shifting gears to carbon intensity of primary energy. Carbon Intensity refers to the rate of emission for a specified pollutant in relation to the intensity of a defined activity or industrial operation. For primary energy, carbon intensity is also known as emission factor, denoted in terms of unit weight of carbon dioxide ($CO_2$) emitted per unit energy generated. Emissions of $CO_2$ resulting from fuel combustion can be accurately estimated irrespective of the utilization method of the fuel, as these emissions are predominantly contingent on the fuel's carbon composition.

Regarding the carbon intensity of each energy sources, fossil fuels are the most carbon-intensive sources of primary energy [12]. Among fossil fuels, coal exhibits the highest carbon intensity. This is due to its composition, which is primarily carbon. When burned, every atom of carbon in the coal turns into a molecule of $CO_2$, making coal the most carbon-intensive fossil fuel. In the U.S., for example, burning coal emits around 2.2 kg of $CO_2$ per kilowatt-hour (kWh) of energy produced.

Oil is the next emission intensive primary energy. While its carbon intensity can vary based on its specific composition and the refining process used, on average, it is lower

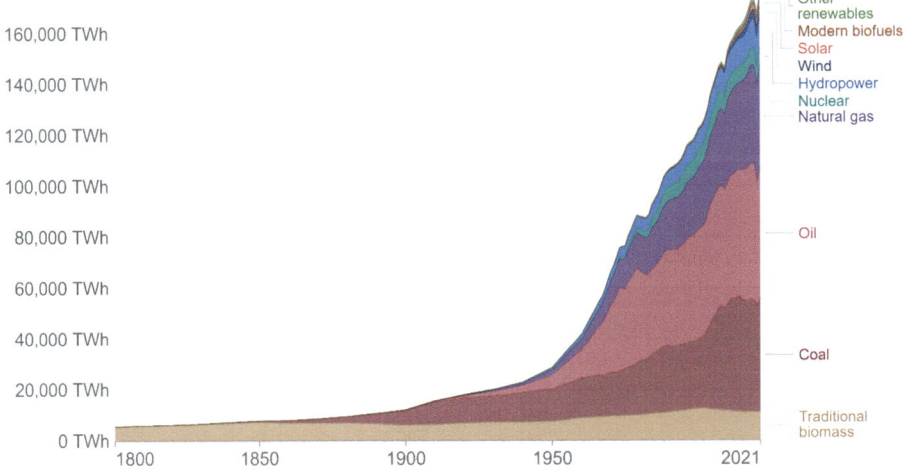

**Fig. 3.2** Historical global primary energy consumption by source

than coal but still considerably high. In general, the combustion of oil emits approximately 1.7 kg of $CO_2$ per kWh of energy.

Natural gas exhibits the least carbon intensity among fossil fuels. It consists mainly of methane, a molecule with one carbon and four hydrogen atoms. When combusted, in addition to $CO_2$, water vapor is produced, which results in a higher energy yield per $CO_2$ molecule emitted. Thus, natural gas produces around 1.0 kg of $CO_2$ per kWh, making it around 50% less carbon-intensive than coal.

Turning to renewable primary energies, nuclear, hydro, wind, and solar energy, are typically considered to have near-zero carbon intensity. They do not involve the combustion of carbon-based fuels, and thus do not directly emit $CO_2$. However, when considering their entire lifecycle—from the extraction and processing of raw materials to the construction and operation of power plants—there can be some indirect emissions. Yet, these emissions are considerably lower than those from fossil fuels. It's also worth noting that these indirect emissions can vary significantly between countries and technologies. Biomass can be a low or high carbon-intensity source, depending on the specific circumstances. If the biomass is sustainably sourced and the growth and harvesting of the biomass recapture the $CO_2$ emitted when the biomass is burned, it could be considered carbon neutral. However, if unsustainable practices are used (for example, clear-cutting forests for biomass production), the carbon intensity could be high.

Figure 3.3 shows the global $CO_2$ emissions by different fuel consumptions. The trend of carbon intensity for various fuel types from 1750 to 2021 shows a gradual evolution

## 3.3 Grid Emission Factor

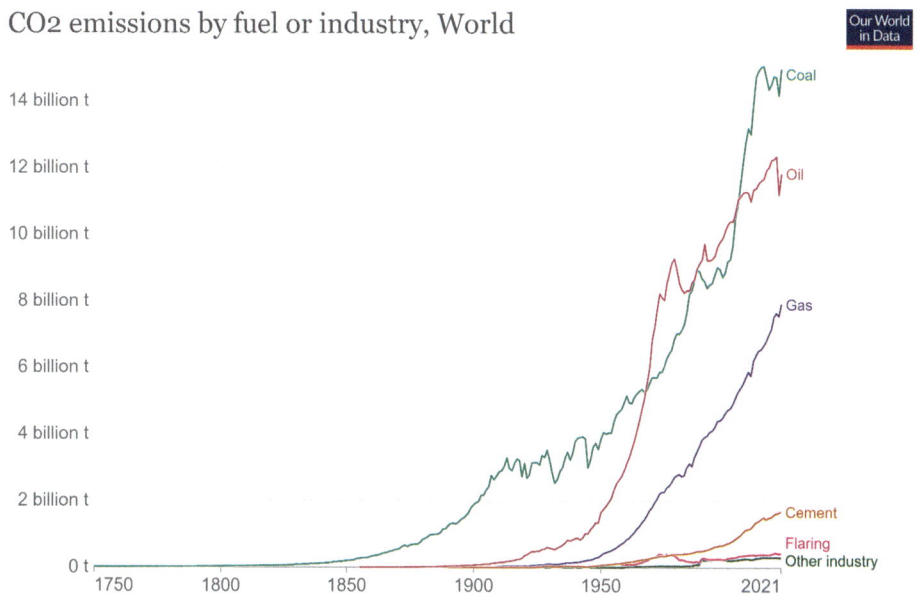

**Fig. 3.3** CO$_2$ emissions for different fuels

in line with technological advancements, resource availability, and environmental consciousness. From a macroscopic perspective, it can be observed that the initial phase of industrialization was predominantly driven by the exploitation of solid fuels—a fact that becomes prominently visible upon altering the data representation in the chart to a relative viewpoint. Industrial-scale power generation via coal combustion first surfaced in Europe and North America during the eighteenth century. It wasn't until the closing decades of the nineteenth century that a surge in emissions attributable to oil and gas production was registered. It took another century before emissions linked to flaring (burning off associated gas during petroleum extraction) and cement production began to manifest. In the contemporary era, energy production is primarily dictated by solid and liquid fuels, though contributions from gas production are also noteworthy. From a global standpoint, the roles of cement manufacturing and flaring remain relatively minor in comparison.

## 3.3 Grid Emission Factor

Regarding the carbon intensity of electricity, it is also referred as the grid emission factor. It is defined as the amount of $CO_2$ emission for each unit of electricity generated. A literature review [13] of total life cycle energy sources $CO_2$ emissions per unit of generated electricity shows the grid emission factor as shown in Table 3.2.

**Table 3.2** Grid emission factor for major primary source to generate electricity [13]

| Primary energy source | Emission factor ($gCO_2$-eq/kWh) |
|---|---|
| Hydroelectric | 4 |
| Wind | 12 |
| Nuclear | 16 |
| Biomass | 230 |
| Solar thermal | 22 |
| Geothermal | 45 |
| Solar (PV) | 46 |
| Natural gas | 469 |
| Coal | 1001 |

Figure 3.4 shows the global carbon intensity of electricity in 2022. The carbon intensity of electricity can vary considerably from region to region, depending on the mix of energy sources used for electricity generation. Here's a general overview as of my training data up until September 2021:

**North America**: In the United States and Canada, the carbon intensity of electricity has traditionally been high due to the heavy reliance on coal and natural gas. However, these

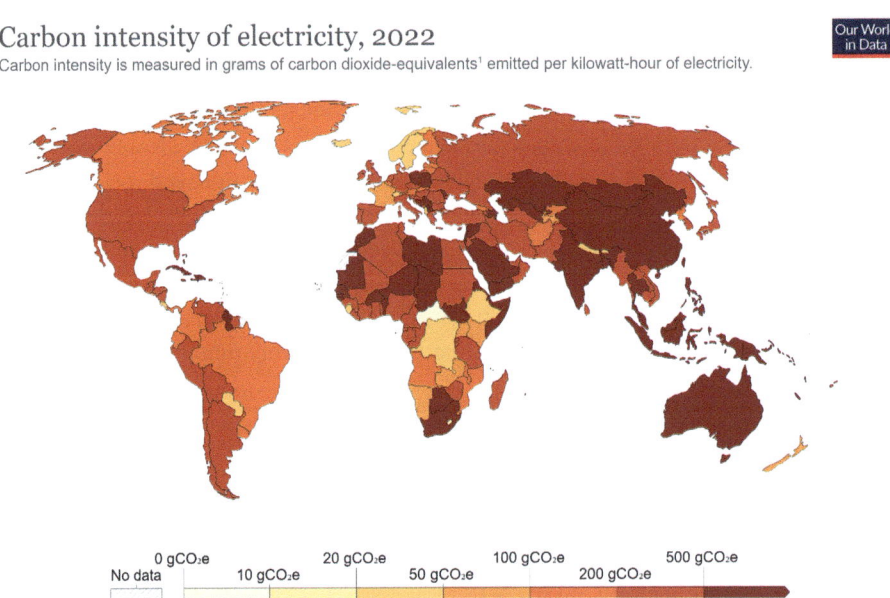

**Fig. 3.4** Global carbon intensity of electricity in 2022

countries have seen a significant reduction in carbon intensity in recent years, due to an increasing share of renewables in the energy mix and a move away from coal towards natural gas, which has lower carbon intensity.

**Europe**: European countries have varied carbon intensities, largely dependent on the specific energy policies of each nation. Countries like Norway, Sweden, and Iceland have very low carbon intensity due to extensive use of hydroelectric and geothermal energy. On the other hand, countries like Poland, which relies heavily on coal, have higher carbon intensity. However, the European Union's aggressive climate goals are driving a transition towards renewable energy across the continent.

**Asia**: China and India, the two most populous countries, have high carbon intensities due to their dependence on coal for electricity generation. However, both countries are investing heavily in renewable energy, which should reduce their carbon intensities in the future. Japan, following the Fukushima disaster, has relied more on fossil fuels, thus increasing its carbon intensity. However, the government has plans to ramp up renewable energy to reduce carbon intensity.

**Africa**: Africa's carbon intensity is varied. Some countries, like South Africa, have high carbon intensity due to their heavy reliance on coal. However, many African countries have low electricity consumption and carbon intensity because a large percentage of their population lacks access to electricity. Africa has significant renewable energy potential, and its carbon intensity could remain low if this potential is realized.

**South America**: South America generally has lower carbon intensity compared to global averages. Countries like Brazil and Paraguay have extensive hydroelectric power, leading to low carbon intensity. However, there are concerns about the environmental impacts of large-scale hydroelectric projects.

**Australia**: Australia has one of the highest carbon intensities among developed countries due to its heavy reliance on coal for electricity generation. However, there are efforts to transition towards renewable energy, which should reduce the carbon intensity in the future.

In conclusion, carbon intensity is influenced by a range of factors, including the availability of natural resources, national energy policies, technological development, and economic considerations. Therefore, the transition to low carbon intensity electricity generation is not only a technical challenge but also a socio-economic one. As the world moves towards decarbonization, regions must take a comprehensive and balanced approach to achieve a sustainable energy future. Please note that the data and trends may have changed after my training cut-off in September 2021.

Copyright for all figures: Please refer to https://ourworldindata.org/, the authors have explicitly made the claims (copied as below without any change).

# Calculation Approach and Assumptions

# 4

Carbon emissions associated with operational energy of each thermal system are derived for all technologies discussed in the previous chapter. Governing equations for individual system categories are provided in the following sections. For each case study, three key parameters are calculated: (i) actual thermal load representing the required total primary energy, termed as $Load_{TH}$; (ii) energy efficiency of the thermal system, $\eta_{TH}$; (iii) carbon dioxide equivalent emissions of the energy system and its load, termed as $CO2_{TH}$.

Since onsite power generation technologies are also considered (self-powered, combined heat and power), comparison of thermal systems was conducted based on the carbon footprint associated with both electrical and thermal energy loads by considering the carbon intensity of electrical power supplied. Electrical energy storage devices are not considered in this analysis and any difference between load and demand was directly exchanged with the grid. Additionally, several assumptions are made for analyzing all the energy configurations considered, as shown in Table 4.1. Carbon dioxide footprint of each configuration was calculated based on the primary energy source utilized, i.e., carbon intensity of the electrical grid, hydrogen production carbon intensity, and the $CO_2$ produced via natural gas combustion in the burner. The electrical load was assumed to be 30 kWh per day, while the heat demand was assumed to be 100 kWh per day. The electrical demand chosen here is typical for an average residential building; however, the combined space and water heating energy consumption varies significantly across the seasons and regions. COP of the heat pump is the ratio of useful heating provided to work required by the compressor.

**Table 4.1** Assumptions and operational conditions assumed in the energy and carbon footprint analysis presented

| Thermal technology | Parameter | Value | Units | References |
|---|---|---|---|---|
| All | $Load_{TH}$—building thermal load | 100 (buildings) 10,000 (industry) | kWh/day | Baseline for comparison |
| All | $Load_{EL}$—building electrical load | 30 (buildings) 5,000 (industry) | kWh/day | Baseline for comparison |
| Conventional ER, heat | $\eta_{ER}$—energy efficiency of the electrical resistance heater | 100% | % | |
| Conventional EHP, heat | $\eta_{EHP}$—energy efficiency of the electrical heat pump | COP 3, 300% | COP, % | |
| Conventional FFD, heat | $\eta_{FFD}$—energy efficiency of the fuel fired heater | 95% | % | |
| Gas sorption HP, heat | $\eta_{SorptionHP}$—energy efficiency of the thermally driven sorption heat pump | 150% | % | |
| All | $CO2_{grid}$—carbon intensity of the electrical grid | 10–1000 | g/kWh | |
| All | $CO2_{fuel}$—carbon intensity of the fuel utilized | 10–400 | g/kWh | |
| All | $\kappa$—balance of plant parasitic power (% of thermal load) | 3% | % | [15, 16] |
| SPD, SPD-HP | $\alpha$—fraction of electric energy dissipated as useful heat | 0.5 | | [15] |
| SPD, SPD-HP | $f_{PC}$—fuel supply to power cycle | 1 | | [15, 17] |
| SPD, SPD-HP | $\eta_{PC}$—power cycle conversion efficiency for self-powered devices | 0.1 | | [15, 18] |
| DFHP | $x_{el}$—fraction of electrical driven operational hours in a dual fuel configuration | 60% | % | |
| DFHP | $x_{fuel}$—fraction of fuel driven operational hours in a dual fuel configuration | 40% | % | |
| DFHP | $COP_h$—heat pump's heating COP | COP 3, 300% | COP, % | |

(continued)

## 4.1 Heating Technologies

**Table 4.1** (continued)

| Thermal technology | Parameter | Value | Units | References |
|---|---|---|---|---|
| CHP systems | $\gamma_{heat\,recovery}$—useful heat recovery from CHP based configurations | 90% | % | |
| CHP systems | $t_{operation}$—operational hours of CHP per day | 5–24 | Hours | |
| CHP systems | $\eta_{chp,EL}$—electrical efficiency of the combined heat and power system | 30–50% | % of fuel HHV | |
| CHP systems with TES | $kW_{chp\,TES}$—power rating of the CHP in thermal energy storage configuration | 2–10 (buildings) 200 kW (industry) | kW | [19, 20] |
| BSHP | $kW_{EL,BSHP}$—electrical power rating of the prime mover for black start heat pump | | | |
| BSHP, SFFHP | $COP_h$—heat pump's heating COP | COP 4, 400% | COP, % | [21, 22] |
| Conventional AC | $\eta_{convAC}$—cooling energy efficiency of conventional air conditioner | 300% | % | |
| Gas sorption AC | $\eta_{sorptionAC}$—cooling energy efficiency of thermally driven sorption heat pump | 70% | % | |
| CHP with conventional AC | $CHP_{EL,ConvAC}$—power rating of the CHP in cooling applications | 2 | kW | |
| BSAC | $COP_c$—electric AC's cooling efficiency | 3 | COP | |

## 4.1 Heating Technologies

Required primary energy for supporting the thermal load is calculated according to Eq. 4.1, based on the thermal efficiency of the individual system, calculated according to equations presented in the following subsections.

$$Load_{TH,true} = \frac{Load_{TH}}{\eta_{Thermal\,technology}} \quad (4.1)$$

Carbon emissions are calculated for all configurations according to Eqs. 4.2a and 4.2b. Equation 4.2a represents the carbon footprint of the thermal load while Eq. 4.2b accounts for both thermal and electrical energy related carbon emissions.

$$CO2_{TH} = \frac{\left(\left(Load_{TH,true} \times CO2_{grid\,or\,fuel}\right) + \left(\kappa \times Load_{TH} \times CO2_{grid}\right)\right)}{Load_{TH}} \quad (4.2a)$$

$$CO2_{TH+EL} = CO2_{TH} + CO2_{grid} \quad (4.2b)$$

### 4.1.1 Conventional Electrical Resistance and Heat Pump Systems, Direct Fuel Fired and Thermally Driven Sorption Systems

Equations 4.2a and 4.2b calculates carbon emissions associated with (a) electrical systems including resistance heaters and heat pump, (b) fuel fired as well as fuel driven sorption heat pumps. The balance of plant power is accounted for using a factor $\kappa$, as a percentage of thermal load. Primary energy's carbon footprint is selected depending on the thermal system category, i.e., electrical or fuel-based system.

### 4.1.2 Self-powered Devices

Equations 4.3a and 4.3b are used for calculating the energy efficiency of the fuel based self-powered direct heating and self-powered electrical heat pump system, according to our previous work [15].

$$\eta_{SPD} = \frac{\eta_{FFD}}{(1 + \kappa \times (1 - \alpha))} \quad (4.3a)$$

$$\eta_{SPD-HP} = \frac{\eta_{FFD} \times (1 + ((\eta_{PC} \times f_{PC}) \times (COP_h - 1)))}{(1 + (\kappa \times (COP_h - \alpha)))} \quad (4.3b)$$

### 4.1.3 Dual-Fuel Devices

Equation 4.4 calculates the combined energy efficiency when utilizing both electricity or fuel as primary energy resources in dual fuel systems consisting of fuel fired and electric heat pump hybrid configuration. The term $\alpha$ represents the fraction of electric energy

## 4.1 Heating Technologies

dissipated as useful heat in the hybrid heating system whereas $x_{el}$ and $x_{fuel}$ represent fraction of electrical and fuel driven operational hours, respectively.

$$\eta_{DFHP,SFFHP} = \left( \frac{\left( \frac{x_{el} \times Load_{TH} \times CO2_{grid}}{COP_h} \right) + \left( \frac{x_{fuel} \times Load_{TH} \times CO2_{fuel}}{\eta_{FFD}} \right) + (\alpha \times Load_{TH} \times CO2_{grid})}{Load_{TH}} \right) \quad (4.4)$$

### 4.1.4 CHP Systems with and Without Thermal Energy Storage (TES)

Equations 4.5–4.10 determines the carbon footprint associated with combined heat and power-based energy solutions for buildings. CHP configurations with and without thermal energy storage are considered under this category where any excess electrical energy generated onsite was assumed to be exported back to the grid. Thermal energy storage utilization becomes a requirement for configurations employing continuously operated prime movers (e.g., fuel cells) whereas for prime movers designed for following the thermal load demand (e.g., engine-based configurations), thermal energy storage is less critical when considering space heating as the primary thermal load.

The power rating of the CHP's prime mover in configurations without TES is calculated per the thermal load demand where the prime mover was assumed to be operated in thermal load following mode of operation. Equation 4.5 identifies the thermal power rating of the CHP ($CHP_{TH\ rating}$) by considering the waste heat recovery ($\gamma_{heat\ recovery}$) and cumulative operational hours in a 24-h time period, denoted as $t_{operation}$. Equation 4.6 then calculates the total electrical energy generated ($CHP_{EL,gen}$) in the 24-h period by considering the electrical efficiency of the prime mover ($\eta_{chp,\ EL}$). Excess electrical energy ($\beta_{EL,exp}$) is then calculated according to Eq. 4.7. Equation 4.8 finalizes the carbon intensity of this configuration by accounting for emissions associated with grid export.

$$CHP_{TH\ rating} = \left( \frac{Load_{TH}}{\gamma_{heat\ recovery}} \right) \times \frac{1}{t_{operation}} \quad (4.5)$$

$$CHP_{EL,gen} = \frac{CHP_{TH\ rating} \times t_{operation}}{\left( \frac{1}{\eta_{chp,EL}} - 1 \right)} \quad (4.6)$$

$$\beta_{EL,exp} = CHP_{EL,gen} - Load_{EL} - \kappa \times Load_{TH} \quad (4.7)$$

$$CO2_{chp} = \left[ \frac{\left( \frac{CHP_{EL,gen} \times CO2_{fuel}}{\eta_{chp,EL}} \right) - (\beta_{EL,exp} \times CO2_{grid})}{(Load_{TH} + Load_{EL} + \kappa \times Load_{TH})} \right] \quad (4.8)$$

$$CHP_{TH,TES} = \left(\frac{kW_{chp\,TES}}{\eta_{chp,EL}} - kW_{chp\,TES}\right) \times \gamma_{heat\,recovery} \times t_{operation}$$

For CHP configurations integrated with thermal energy storage, power rating of the prime mover is assumed rather than calculated, denoted as $kW_{chp\,TES}$. This will allow the minimization of excess electrical exports to the grid and is particularly suitable for prime movers that operate in a continuous manner ($t_{operation} = 24$ h), for instance, solid oxide fuel cells (SOFCs). The term $v_{EL,exp,TES}$ in Eq. 4.9 determines the electrical energy exported to the grid while Eq. 4.10 calculates the carbon intensity of CHP systems with integrated thermal energy storage in meeting both electrical and thermal load demand of a building.

$$v_{EL,exp,TES} = kW_{chp\,TES} \times t_{operation} - Load_{EL} - \kappa \times Load_{TH} \quad (4.9)$$

$$CO2_{chp\,TES} = \left[\frac{\left(\frac{kW_{chp\,TES} \times t_{operation} \times CO2_{fuel}}{\eta_{chp,EL}}\right) - \left(v_{EL,exp,TES} \times CO2_{grid}\right)}{(Load_{TH} + Load_{EL} + \kappa \times Load_{TH})}\right] \quad (4.10)$$

### 4.1.5 CHP Systems with Sorption Heat Pump

For CHP configurations integrated with sorption heat pumps, the carbon intensity is calculated according to Eqs. 4.5–4.8 where the thermal rating of the CHP ($CHP_{TH\,rating}$) accounts for higher energy efficiency of the sorption heat pump, $\eta_{sorption\,HP}$. In Eq. 4.5, the term $Load_{TH}$ is replaced with $Load_{TH,\,true}$, defined according to Eq. 4.1.

### 4.1.6 Blackstart Heat Pump Systems with and Without Thermal Energy Storage

Blackstart heat pump systems with and without thermal energy storage are also analyzed with two different prime mover technologies suitable for intermittent/on-demand or continuous operation, for example, using reciprocating engines or fuel cells, respectively. $kW_{EL,BSHP}$ denotes the electrical power rating of the prime mover utilized in the black start system and is calculated according to Eq. 4.11. Equation 4.16 is derived by solving Eqs. 4.12–4.16, as described below.

Since no power generation is needed to support onsite electrical loads at the building or export to the grid, Eqs. 4.12–4.16 are utilized to derive the electrical power rating of the prime mover. Considering the electrical efficiency and heat recovery efficiency, useful thermal power generated by the prime mover is calculated with Eq. 4.12. Total thermal energy needed from the black start heat pump after heat recovery from the prime mover

## 4.2 Cooling Technologies

($Load_{TH,BSHP}$) is calculated according to Eq. 4.13. Actual thermal energy needed by the black start heat pump after heat recovery from the prime mover ($Load_{TH,BSHP,true}$) is calculated per Eq. 4.14 by considering the heating COP ($COP_h$) of the heat pump. Equation 4.15 calculates the electrical energy generated by the prime mover (($CHP_{EL,BSHP}$). Equation 4.16 represents the electrical energy needed by the heat pump and is equal to the electrical energy generated by the prime mover.

$$kW_{EL,BSHP} = \frac{Load_{TH} \times \eta_{chp,EL}}{t_{operation} \times (\eta_{chp,EL} \times COP_h - \eta_{chp,EL} \times \gamma_{heat\,recovery} + \gamma_{heat\,recovery})} \quad (4.11)$$

by solving Eqs. 4.12-4.16

$$kW_{TH,BSHP} = \left(\frac{kW_{EL,BSHP}}{\eta_{chp,EL}} - kW_{EL,BSHP}\right) \times \gamma_{heat\,recovery} \quad (4.12)$$

$$Load_{TH,BSHP} = Load_{TH} - kW_{TH,BSHP} \times t_{operation} \quad (4.13)$$

$$Load_{TH,BSHP,true} = \frac{Load_{TH,BSHP}}{COP_h} \quad (4.14)$$

$$CHP_{EL,BSHP} = kW_{EL,BSHP} \times t_{operation} \quad (4.15)$$

$$CHP_{EL,BSHP} = Load_{TH,BSHP,true} \quad (4.16)$$

Total equivalent carbon dioxide emissions associated with building thermal load provided by black start heat pump configuration ($CO2_{TH,BSHP}$) are calculated according to Eq. 4.17

$$CO2_{TH,BSHP} = \left[\frac{\left(\frac{kW_{EL,BSHP}}{\eta_{chp,EL}} \times t_{operation} \times CO2_{fuel}\right)}{(Load_{TH})}\right] \quad (4.17)$$

## 4.2 Cooling Technologies

Carbon emissions associated with different cooling technologies is calculated from Eqs. 4.1 and 4.2a. The actual building thermal load ($Load_{TH}$) and energy efficiency of the cooling device ($\eta_{Thermal\,technology}$) is considered in calculating the carbon emissions.

## 4.2.1 Conventional and Sorption Air Conditioners

For conventional and thermally driven sorption air conditioners, primary energy's carbon footprint ($CO2_{grid\ or\ fuel}$) is selected depending on the thermal system category being analyzed, i.e., electrical or fuel-based system.

$$Load_{TH,true} = \frac{Load_{TH}}{\eta_{Thermal\ technology}} \qquad (4.1)$$

$$CO2_{TH} = \frac{\left((Load_{TH,true} \times CO2_{grid\ or\ fuel}) + (\kappa \times Load_{TH} \times CO2_{grid})\right)}{Load_{TH}} \qquad (4.2a)$$

$$CO2_{TH+EL} = CO2_{TH} + CO2_{grid} \qquad (4.2b)$$

## 4.2.2 CHP Configurations with Conventional AC or Sorption AC

For conventional and sorption air conditioning systems integrated with combined heat and power systems, carbon emissions are calculated as below.

Total equivalent carbon dioxide emissions associated with building thermal and electrical load installed with CHP based conventional AC configurations are calculated from Eq. 4.19 by considering electrical AC's cooling efficiency ($COP_C$) and assumed electrical power rating of the ($CHP_{EL,\ conv\ AC}$).

$$CO2_{CHP\ conv\ AC} = \frac{\left(\frac{CHP_{EL,\ Conv\ AC} \times t_{operation} \times CO2_{fuel}}{\eta_{chp,EL}}\right) - \left(\left(CHP_{EL,\ Conv\ AC} \times t_{operation} - \left(\frac{Load_{TH}}{COP_C}\right) - Load_{EL} - \kappa \times Load_{TH}\right) * CO2_{grid}\right)}{(Load_{TH} + Load_{EL} + \kappa \times Load_{TH})} \qquad (4.19)$$

In CHP systems integrated with sorption-based air conditioners, the thermal power rating needed from the CHP is calculated per Eq. 4.5. Electrical energy generated by the CHP is calculated according to Eq. 4.6. Excess electrical energy generation is assumed as exported to the power grid and is calculated according to Eq. 4.7, similar to the approach adopted with heating technologies.

$$CHP_{THrating} = \left(\frac{Load_{TH,true}}{\gamma_{heat\ recovery}}\right) \times \frac{1}{t_{operation}} \qquad (4.5)$$

$$CHP_{EL,gen} = \frac{CHP_{THrating} \times t_{operation}}{\left(\frac{1}{\eta_{chp,EL}} - 1\right)} \qquad (4.6)$$

$$\beta_{EL,exp} = CHP_{EL,gen} - Load_{EL} - \kappa \times Load_{TH} \qquad (4.7)$$

## 4.2 Cooling Technologies

Equation 4.20 is used for calculating the carbon emissions from CHP configurations integrated with sorption-based AC systems.

$$CO2_{chp\,sorptionAC} = \left[ \frac{\left(\frac{CHP_{EL,gen} \times CO2_{fuel}}{\eta_{chp,EL}}\right) - \left(\beta_{EL,exp} \times CO2_{grid}\right)}{(Load_{TH} + Load_{EL} + \kappa \times Load_{TH})} \right] \quad (4.20)$$

### 4.2.2.1 Blackstart AC Systems with Engine or Fuel Cell as PM

The calculation approach for blackstart conventional AC systems is similar to the one utilized in blackstart heat pump heating technologies, however since the waste heat from CHP is not used in the cooling applications, the heat recovery ratio is set as zero in Eq. 4.11, yielding Eq. 4.21. As shown, Eq. 4.21 calculates the required electrical power rating of the prime mover for black start AC systems.

$$kW_{EL,BSAC} = \frac{Load_{TH}}{t_{operation} \times COP_c} (4.11 \; with \; \gamma_{heat\,recovery}=0) \quad (4.21)$$

Total equivalent carbon dioxide emissions associated with building thermal and electrical load installed with based black-start AC configurations.

$$CO2_{TH,BSAC} = \left[ \frac{\left(\frac{kW_{EL,BSAC}}{\eta_{chp,EL}} \times t_{operation} \times CO2_{fuel}\right)}{(Load_{TH})} \right] \quad (4.22)$$

# Effective Carbon Footprint of Heating Technologies

# 5

Recent global energy review by International Energy Agency (IEA) for the year 2021 concluded that the emissions from the world's power plants reached their highest level in the history. Additionally, the highest increase of 900 Mt in $CO_2$ emissions by sector took place in electricity and heat production [23]. Since the two primary energy resources utilized in meeting this demand are fuel and electricity, all possible configurations utilized in providing the required energy in buildings and industry are analyzed. As discussed in Chap. 3, carbon intensity of global electricity production ranges from 10 to 1,200 $gCO_{2,eq}$/kWh [24]. Similarly, direct combustion of fuel for the same thermal energy ranges from 10 to 400 $gCO_{2,eq}$/kWh [25]. For the comparative analysis provided below, a combination of primary energy utilization possibilities within the whole range of carbon intensities were considered.

## 5.1 Carbon Footprint of Non-power Generating Thermal Technologies

Thermal technologies which convert the supplied primary energy directly to heat energy are considered here. Conventional heating technologies employed in building heating typically consist of one of: electric resistance (ER), electric heat pump (EHP), fuel fired device (FFD). Configurations such as dual fuel heat pump (DFHP) and sorption heat pump (SHP) are not commonplace but are becoming a preferable choice of commercial building customers. Seamlessly fuel flexible heat pump (SFFHP) is a brand-new technology that has been demonstrated to offer better efficiency than any of the existing dual fuel heat pumps. As a result, a comparison of the carbon footprint of all these technologies

© The Author(s), under exclusive license to Springer Nature Switzerland AG 2024
P. Cheekatamarla et al., *The Role of Fuels in Transforming Energy End-Use in Buildings and Industrial Processes*, Synthesis Lectures on Engineering, Science, and Technology,
https://doi.org/10.1007/978-3-031-45365-6_5

**Table 5.1** Carbon intensity of thermal load provided by non-power generating heating technologies supplied with different electric grid carbon intensity factors. Natural gas as the fuel based primary energy

| No Onsite Power Generation | Carbon intensity of thermal energy with different electric grid carbon intensity factors - natural gas for fuel systems | | | | | | |
|---|---|---|---|---|---|---|---|
| | 1200/180 | 1000/180 | 800/180 | 600/180 | 400/180 | 200/180 | 10/180 |
| Conv ER - 100% | 1236 | 1030 | 824 | 618 | 412 | 206 | 10 |
| Conv EHP - 300% | 436 | 363 | 291 | 218 | 145 | 73 | 4 |
| DFHP (60/40) - 220% | 352 | 306 | 260 | 214 | 168 | 122 | 78 |
| SFFHP - 400% | 336 | 280 | 224 | 168 | 112 | 56 | 3 |
| Sorption HP - 150% | 156 | 150 | 144 | 138 | 132 | 126 | 120 |
| Conv FFD - 95% | 225 | 219 | 213 | 207 | 201 | 195 | 190 |

in meeting the thermal needs of a building consuming 100 kWh of thermal energy daily, calculated according to the equations (4.1)–(4.22) presented in Chap. 4 was conducted.

Table 5.1 displays the heating technology and its assumed energy efficiency. Each column's header displays the carbon intensity of the primary energy, where the first value represents the electrical grid's number while the second one is that of the natural gas. As shown, the electrical resistance heater with 100% energy efficiency has the highest carbon footprint compared to all other technologies except when the electrical grid carbon intensity is near zero. However, natural gas-based technologies are favorable when the grid's carbon intensity is >400 gCO$_{2e}$/kWh.

Considering all the technologies but electrical resistance heater, as shown in Fig. 5.1, SFFHP with an equivalent energy efficiency of 400% (COP—4) followed by SHP (fuel driven thermal sorption heat pump, equivalent energy efficiency of 150%, COP—1.5) remain favorable options at moderate and high electrical grid carbon intensities. SFFHP and conventional heat pump are the best decarbonizing technologies at electric carbon intensities <200 gCO$_{2e}$/kWh. Considering the observed trends, it is important that the carbon intensity of the fuel is also taken into account, given the employment of different fossil fuels across the globe, ranging from oil to biogas where the carbon footprint ranges from 400 to 10 gCO$_{2e}$/kWh. Hence, the carbon intensity of the same 100 kWh/day thermal load supported by the six basic technologies was calculated at different fuel carbon intensities, as displayed in Fig. 5.2. Also, the electric grid carbon intensity was assumed to be 700 gCO$_{2e}$/kWh, representing global regions with polluting electrical grids. Electrical resistance heating is unfavorable in all possible primary energy choices. SHHHP and SHP are the most favorable solutions if the fuel choice has a carbon intensity of >200 gCO$_{2e}$/kWh (e.g., oil heating). SHP and FFD (95% energy efficiency) become highly attractive if the fuel's carbon intensity remains below <100 gCO$_{2e}$/kWh (e.g., hydrogen, biodiesel, biogas, renewable natural gas etc.).

## 5.1 Carbon Footprint of Non-power Generating Thermal Technologies

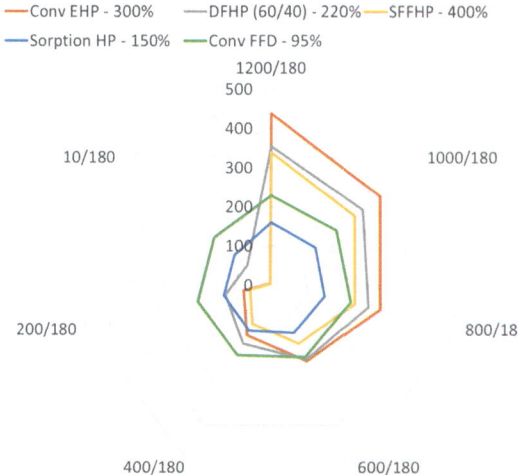

**Fig. 5.1** Carbon footprint associated with thermal loads in a building consuming 100 kWh/day supplied with electrical grid or natural gas fuel

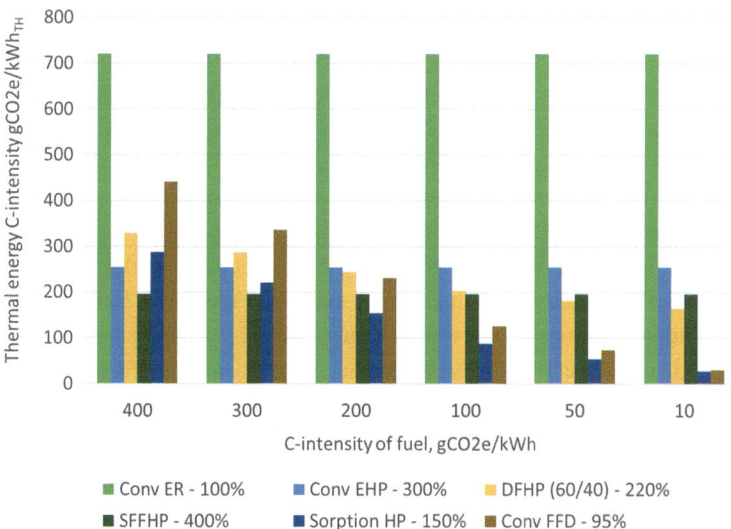

**Fig. 5.2** Carbon footprint associated with thermal loads in a building consuming 100 kWh/day supplied with electrical grid (700 $gCO_{2e}$/kWh) and fuel with different carbon intensities

## 5.2 Carbon Footprint of Power Generating Thermal Technologies

On-site power generation for supporting either the balance-of-plant (BOP) electrical load of the thermal technology or producing excess power beyond the BOP load to supplement building electrical loads are considered for the carbon emissions analysis. Power generation systems employing two different types of prime movers are considered. For instance, a reciprocating engine supplied with fuel can follow the thermal load on-demand, where the prime mover starts producing electrical and thermal energy as soon as the load demand occurs. In these cases, a small-scale energy storage device may be used to minimize the impact of transient loads. Another prime mover technology category includes continuous operation (e.g., a solid oxide fuel cell, thermophotovoltaic etc.) where the PM's kW rating can be optimized for addressing the load profiles. In such cases, energy storage capacity needs to be carefully designed to manage the loads efficiently.

Self-powered devices (SPD) consist of a thermal-to-electric power generation technology integrated with conventional fuel fired devices, similar to the one described in a recent study [16]. Blackstart heat pump systems (BHP) consist of a prime mover technology with defined electrical efficiency that are integrated with or without energy storage to be utilized directly with the heat pump. No excess electricity is produced in this case. Continuous operation of the PM (e.g., a fuel cell system with 50% electrical efficiency) requires electrical and thermal energy storage whereas on-demand systems such as heat engines do not require significant onboard energy storage. Combined heat and power (CHP) systems also utilize the two PM categories described above but are connected directly to the building heating air circulation system either directly (e.g., similar to a forced air system) or indirectly (e.g., via hydronic loop). An additional system considered in this analysis includes CHP system integrated with sorption heat pump where the PM is an on-demand power generator with 33% electrical efficiency. For thermal load following systems such as CHP without thermal energy storage (TES) and CHP with sorption heat pump, 6 operational hours per day was assumed.

Figure 5.3 displays the total equivalent carbon intensity associated with combined electrical and thermal load of the building utilizing fuels with different carbon intensities. All the systems assumed 90% heat recovery from the prime mover. As described in Sect. 5.1, these fuels vary from fossil fuel such as oil to renewable fuels such as biodiesel or biogas. The electrical grid's carbon intensity was assumed to be 700 $gCO_{2e}$/kWh. As shown, CHP systems offer lowest carbon footprint of all power generation-based heating technologies examined. Since these CHP systems also export excess electricity back to the grid, significant carbon savings are possible with low carbon intensity fuel generated electricity versus that generated by the high carbon intensity grid. As a result, negative carbon emissions can be realized with CHP systems operated with clean fuels. Also, BSHP systems are highly desirable if the carbon intensity of fuel is high, for instance above 50 $gCO_{2e}$/kWh.

## 5.2 Carbon Footprint of Power Generating Thermal Technologies

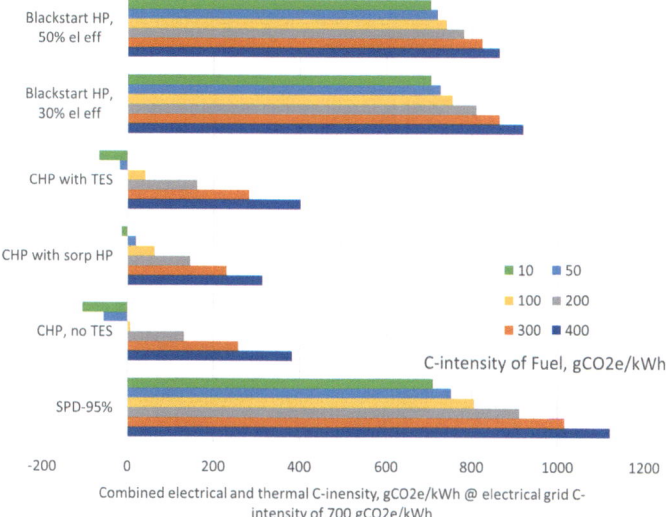

**Fig. 5.3** Combined electrical and thermal load carbon footprint in a building consuming 100 kWh/day thermal energy and 30 kWh/day electrical energy supplied with electrical grid (700 g$CO_{2e}$/kWh) and fuel with different carbon intensities

Figure 5.4 shows the combined electrical and thermal energy related carbon emissions intensity of all onsite power generating thermal technologies in a location where electric grid is extremely clean, generating only 50 g$CO_{2e}$/kWh. In such a location, BSHP systems are generally more favorable compared the CHP systems. However, at fuel carbon intensities below 50 g$CO_{2e}$/kWh, CHP systems are preferred due to higher carbon emissions reduction potential.

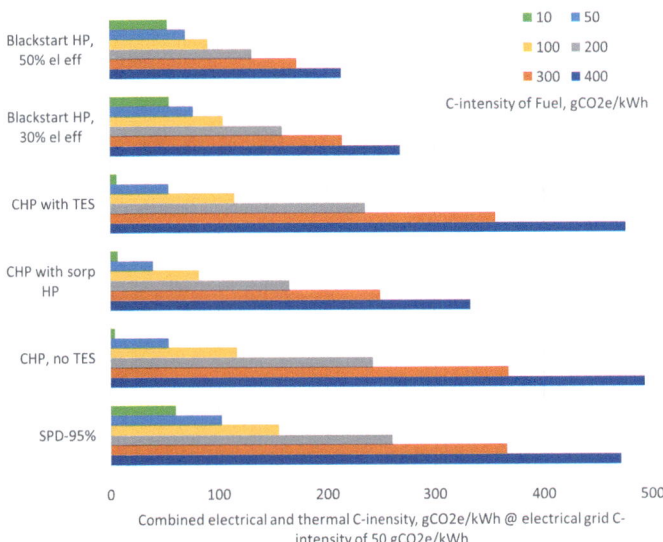

**Fig. 5.4** Combined electrical and thermal load carbon footprint in a building consuming 100 kWh/day thermal energy and 30 kWh/day electrical energy supplied with electrical grid (50 gCO$_{2e}$/kWh) and fuel with different carbon intensities

# 6. Effective Carbon Footprint of Cooling Technologies

Globally, cooling thermal loads are continuously increasing. The demand for air conditioning systems is one of the most critical blind spots in the ongoing energy transition. Space cooling demand accounts for nearly 20% of the total electricity used in buildings around the world today [26]. Rising space cooling energy demand is also straining the electric grid systems globally, leading to higher emissions. The global energy demand associated with space conditioning is anticipated to increase significantly for decades to come, driven by increased affordability and global population growth. This presents a significant opportunity to quickly influence the growth of cooling-related energy demand through policies to improve equipment efficiency. Unlike the wide range of heating system options, cooling systems options are limited and can be broadly categorized into ones with and without onsite power generation.

## 6.1 Carbon Footprint of Non-power Generating Cooling Technologies

Within the non-power generating technology options, conventional air conditioner with 300% electrical efficiency and fuel driven sorption air conditioning systems with 70% energy efficiency are the primary options suitable for buildings. Figure 6.1 shows the total carbon intensity of the daily building cooling load assumed as 100 kWh. The electric grid carbon intensity was assumed as 700 $gCO_{2e}$/kWh, representing a highly polluting utility. As shown, the sorption-based systems are not favorable unless they are operated with fuels having carbon intensity of less than 100 $gCO_{2e}$/kWh. Access to clean, renewable

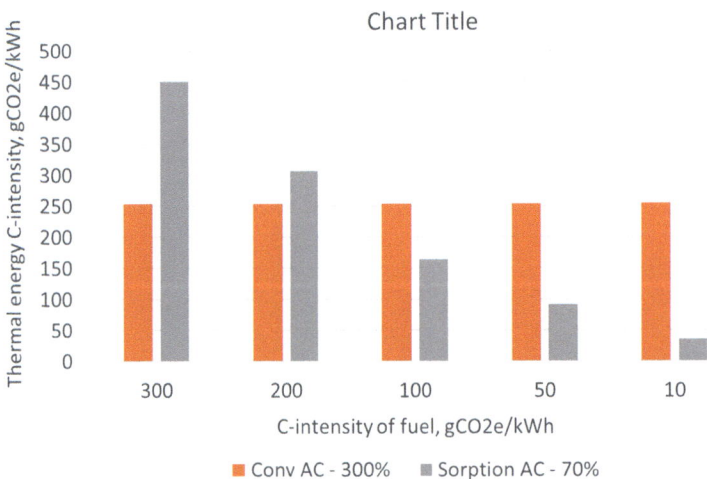

**Fig. 6.1** Carbon footprint associated with cooling load in a building supplied with electrical grid at a carbon intensity of 700 g$CO_{2e}$/kWh and fuels with different carbon intensities

fuels is necessary for thermally driven sorption systems to be attractive in regions with highly polluting grids.

Similarly, the impact of utilizing natural gas fed sorption system in regions served by electrical grid with different carbon intensities in the range of 10–900 g$CO_{2e}$/kWh is shown in Fig. 6.2. As displayed, natural gas based sorption systems are not favorable unless the electric grid is extremely polluting, with carbon intensities greater than 700 g$CO_{2e}$/kWh.

## 6.2 Carbon Footprint of Power Generating Cooling Technologies

Similar to the power generating heating technologies described in Sect. 5.2, four different technologies were considered in cooling applications. Blackstart air conditioners (BSAC) with 50 and 30% electrically efficient prime movers were considered. A 2 kW CHP with 30% efficient prime mover was also considered. In all three cases, the electrical energy produced was assumed to be continuously produced and stored for on-demand usage and the thermal energy was unutilized. For thermally driven sorption-based system, the heat recovery was assumed to be 90% while the electrical efficiency was assumed as 33% in a thermal load following configuration, i.e., the prime mover produces energy on-demand without any major thermal storage considerations. Given the lower energy efficiency of the sorption cooling system at 70%, the total thermal energy demand is much higher than that of a conventional air conditioner. As a result, the PM's power rating is higher and

## 6.2 Carbon Footprint of Power Generating Cooling Technologies

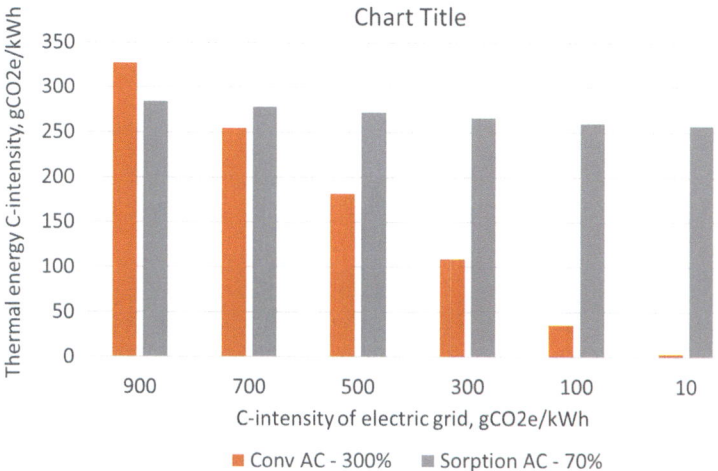

**Fig. 6.2** Carbon footprint associated with cooling load in a building supplied with electrical grid at different carbon intensities, compared with a sorption system fueled by natural gas at a carbon intensity of 180 $gCO_{2e}$/kWh

results in excess electrical power generation which was assumed as exported to the grid after onsite utilization.

As shown in Fig. 6.3, CHP systems are generally favorable in regions with electric grid carbon intensity of greater than 300 $gCO_{2e}$/kWh. Within the CHP configurations analyzed, sorption-based systems have the capability to show lower carbon footprint due to thermal energy utilization. The two case studies investigated here cannot be compared since the power ratings chosen were not optimized. PM's power rating can greatly influence the amount of excess electricity produced and associated grid carbon emissions reduction. As the grid becomes cleaner, black start AC systems are more favorable compared to the CHP systems. In clean grid utility districts however, the conventional air conditioner is a better choice if resiliency is not a major concern.

Figure 6.4 shows the combined electrical and cooling load demand's carbon footprint when serviced by electrical grid with a carbon intensity of 700 $gCO_{2e}$/kWh and fuels with a carbon factor in the range of 300 $gCO_{2e}$/kWh (e.g., oil) to 10 $gCO_{2e}$/kWh (biodiesel, biogas, hydrogen etc.). CHP systems are highly favorable in this scenario since the carbon intensity of electrical power is significantly lower compared to the electrical grid. Onsite thermal energy utilization additionally increases the carbon emissions reduction capability. As explained in Fig. 6.3, excess electrical energy production due to unoptimized PM power rating can significantly impact the overall carbon emissions associated with CHP configurations.

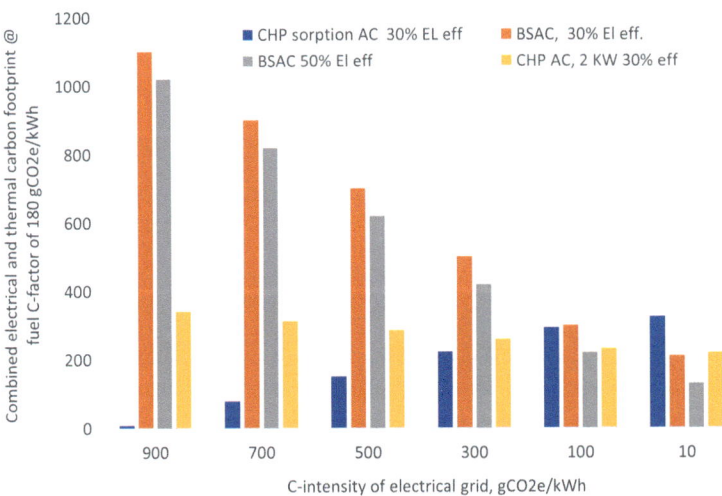

**Fig. 6.3** Combined electrical and cooling thermal load carbon footprint in a building supplied with electrical grid at different carbon intensities and onsite power generation systems fueled by natural gas at a carbon intensity of 180 $gCO_{2e}$/kWh

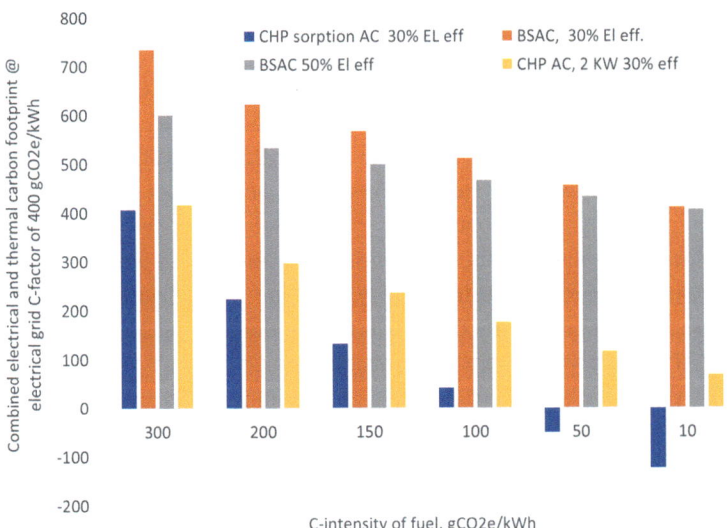

**Fig. 6.4** Combined electrical and cooling thermal load carbon footprint in a building supplied with electrical grid at a carbon intensity of 700 $gCO_{2e}$/kWh and fuels with different carbon intensities

# Effective Carbon Footprint of Industrial Heating Technologies

# 7

Decarbonizing the industrial sector is critical to achieving net-zero emissions, economy-wide. In 2020, the industrial sector accounted for 33% of the nation's primary energy use and 30% of energy-related carbon dioxide ($CO_2$) emissions [27]. However, the industrial sector is considered one of the most difficult to decarbonize due to the diversity and complexity of energy inputs, processes, and operations [28]. Process heating, or thermal processing, is essential to manufacture a wide variety of industrial and consumer products, including those made from metal, plastic, rubber, carbon fiber, concrete, glass, ceramics, and biomass. Thermal energy is needed to transform materials through processes such as drying, curing, melting, forming, sintering, calcining, and smelting [29].

Process heating represents the largest energy use and the largest source of GHG emissions in the manufacturing sector. In 2018, process heating accounted for 31% of sectoral energy use (7.5% of economy-wide energy use) and 51% of sectoral energy-related GHG emissions (10% of economy-wide energy-related GHG emissions) [30]. Direct energy use for process heating in the manufacturing sector includes fuel combustion (66%), steam (30%), and electricity (4%). When the fuel used to generate steam and electricity is considered, at least 95% of energy used for process heat is derived from combustion, including both fossil fuels and waste/byproduct fuels. Furthermore, about one-third of energy consumed in providing process heat is ultimately lost as waste heat.

It is estimated that 30% of industrial heating applications required heat below 100 °C, 27% between 100 and 400 °C, 43% need heat above 400 °C [30]. As a result, industrial heating technologies can be broadly categorized in to processes which require temperatures below 75 °C where emerging high temperature heat pumps can be utilized in addition to conventional heating equipment. For processes requiring higher temperatures, the equipment options are limited and usually involve electrical resistance, gas combustion, and combined heat and power. In the analysis presented below, the daily thermal and electrical load of the industrial complex was assumed to be 10 and 5 MWh in addition to the BOP (for thermal energy movement) load of 0.3 MWh.

## 7.1 Low Temperature Processes

Industrial processes requiring temperatures below 75 °C can utilize technologies similar to the ones described in Sect. 5.1. Conventional fuel fired devices such as boilers, high temperature heat pump, electric resistance heater, thermally driven sorption heat pumps, dual fuel and SFFHP are all possible options. Assumed energy efficiencies for all the devices analyzed are presented in Fig. 7.1 where natural gas was assumed as the fuel while the electric grid carbon intensity was varied between 10 and 900 gCO$_{2e}$/kWh, emulating that of global electric grids discussed in Table 3.2 in Chap. 3. As shown, electrical resistance heating will be favorable under clean electric grid scenarios. EHP, SFFHP and DFHP are the most favorable technologies compared to gas fired technology if the electrical grid carbon intensity is below 700 gCO$_{2e}$/kWh. EHP and SFFHP systems remain favorable compared to sorption heat pumps if the electric grid's carbon intensity is below 500 gCO$_{2e}$/kWh.

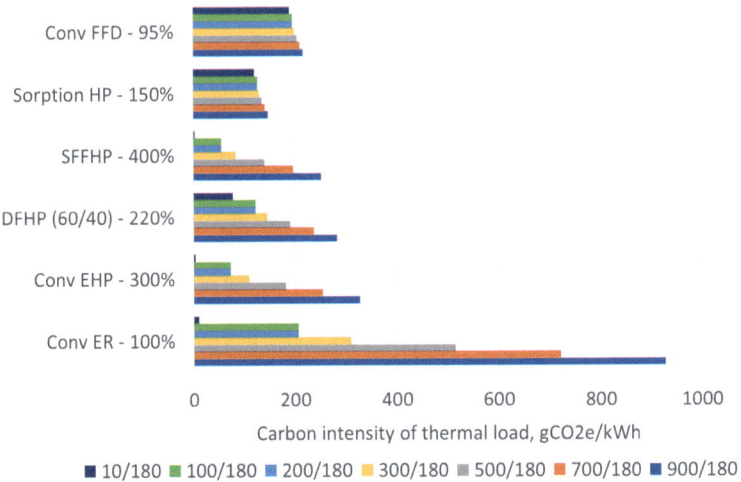

**Fig. 7.1** Carbon intensity of thermal load provided by non-power generating heating technologies supplied with different electric grid carbon intensity factors. Natural gas as the fuel based primary energy

A comparison of the carbon footprint of industrial thermal energy consumption at 10 MWh/day located in two different electrical grid utility districts with high and low carbon intensities is shown in Fig. 7.2. Figure 7.2a represents an industrial customer located in regions serviced by high carbon footprint electrical utility with a carbon intensity of 700 $gCO_{2e}$/kWh while Fig. 7.2b represents a region with low carbon footprint electrical grid with carbon intensity of 50 $gCO_{2e}$/kWh. As shown in Fig. 7.2a, amongst all the technologies examined in industries served by highly polluting grid, the sorption heat pump offers the lowest carbon footprint except when using fuels with carbon intensity of 300 $gCO_{2e}$/kWh. Additionally, if the carbon intensity of the fuel is below 150 $gCO_{2e}$/kWh, a conventional fuel fired device is the second-best option after the sorption heat pump. It is clear that the heat pumps do not offer any benefit in regions with highly polluting electrical grid. Conversely, as shown in Fig. 7.2b, heat pump-based solutions are extremely beneficial, particularly EHP and SFFHP if the carbon intensity of the fuel is above 50 $gCO_{2e}$/kWh. However, the sorption heat pump still remains as the lowest carbon footprint solution if served by a fuel with carbon intensity below 50 $gCO_{2e}$/kWh. In lieu of a heat pump solution, an electrical resistance-based heating system is the preferred solution if the carbon footprint of fuel is above 100 $gCO_{2e}$/kWh.

## 7.2 High Temperature Processes

Many modern industrial processes use electricity or fuel as the primary energy for processes such as annealing, curing, powder coating, drying, laminating, preheating, thermal forming, sintering etc. as shown in Fig. 7.3. Product categories include metals, plastics, glass, rubber, paper, food, fabric, MDF boards, wood, ceramic, etc., where the operating temperature exceeds 200 °C and reaches up to 1200 °C. The primary heating equipment options in such processes include electrical resistance heaters, fuel fired equipment, and combined heat and power systems. Conventional fuel cell, turbine and engine-based CHP systems are limited to temperatures below 700 °C in general. Other emerging technologies such as induction heating, microwave heating are also viable options where the heat transfer efficiency is higher compared to electrical resistance heating.

Figure 7.4 displays the carbon footprint associated with combined electrical and thermal load in an industrial complex consuming 5 MWh and 10 MWh per day respectively serviced by electric grid and/or fuel with different carbon intensities. Figure 7.4a displays the total industrial energy carbon intensity in regions with highly polluting electrical grid at 700 $gCO_{2e}$/kWh. As shown, CHP with integrated thermal energy storage followed by direct fuel fired heating are better solutions when utilizing fuels with wide range of carbon intensities. Similarly, in regions where the carbon footprint of the electrical grid is moderately polluting at 400 $gCO_{2e}$/kWh (Fig. 7.4b), CHP and FFD solutions are still preferable with all fuels except the ones where the fuel source is extremely carbon heavy at 400 $gCO_{2e}$/kWh (e.g., tar, heavy oils etc.). Conversely, in regions serviced by clean electric

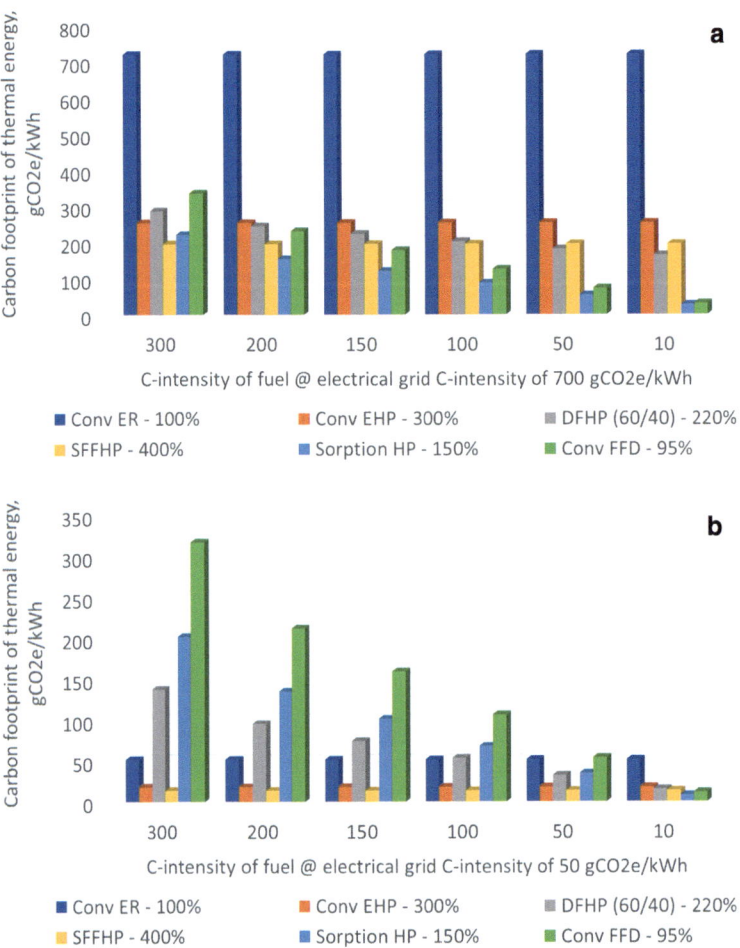

**Fig. 7.2** Carbon footprint of thermal load provided by non-power generating heating technologies supplied with different carbon intensity factors of the fuel supplied and electrical grid with carbon intensity of **a** 700 gCO$_{2e}$/kWh (carbon heavy), **b** 50 gCO$_{2e}$/kWh (clean/renewable)

grid utilities with 50 gCO$_{2e}$/kWh carbon intensity, conventional electrical resistance-based heating shows better environmental benefit compared to the other two technologies if the carbon intensity of the fuel is above 100 gCO$_{2e}$/kWh. CHP and FFD devices are however favorable if the fuel source is clean, for instance at carbon intensities below 50 gCO$_{2e}$/kWh.

## 7.2 High Temperature Processes

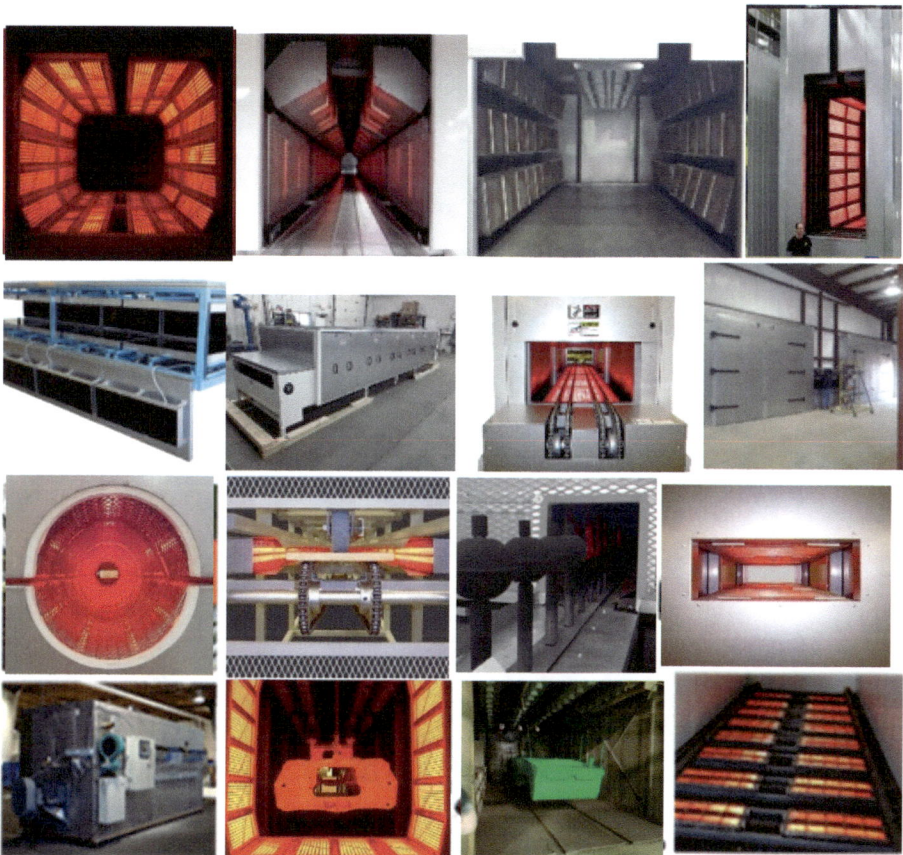

**Fig. 7.3** Industrial heating equipment for processing metals, fabric, plastic, glass, ceramic, food, wood, coatings etc. (with permission from Trimac Industrial Systems)

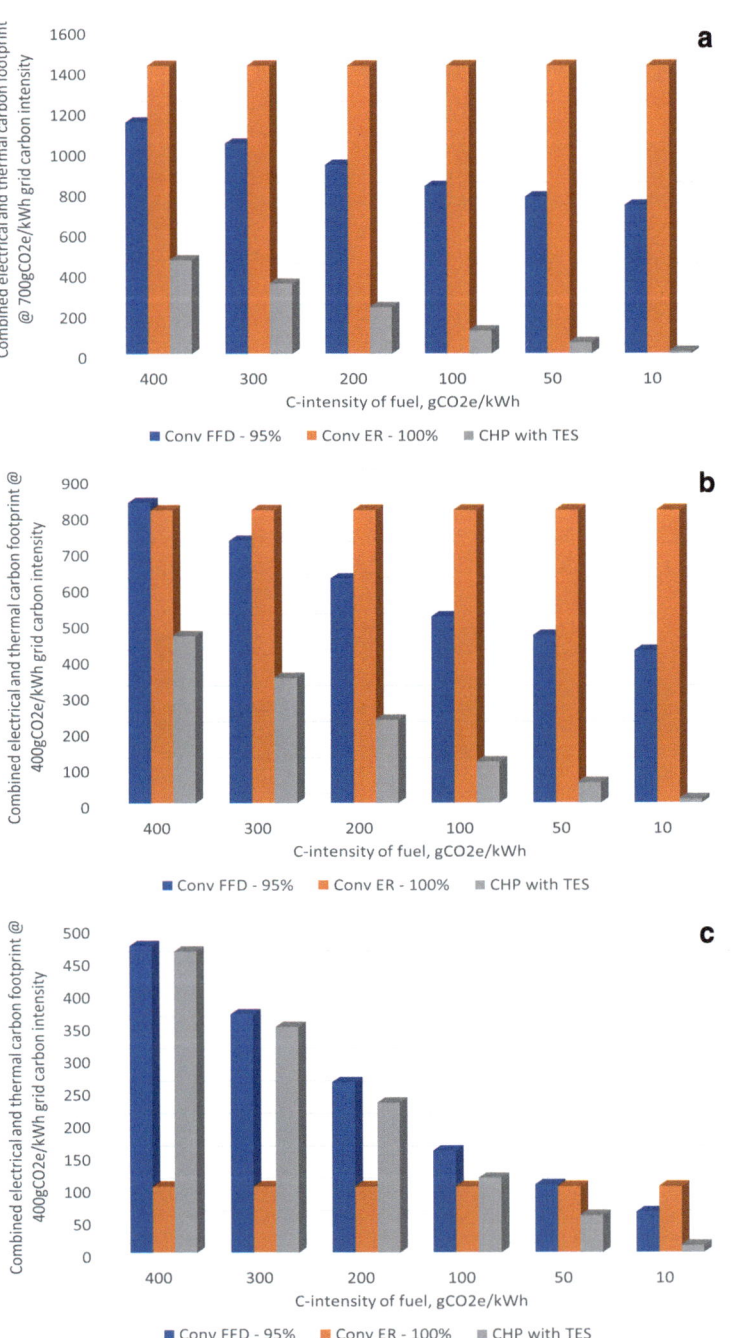

**Fig. 7.4** Carbon footprint of combined thermal and electrical in an industry supplied with different carbon intensity factors of the fuel along with electrical grid at carbon intensity of: **a** 700 $gCO_{2e}$/kWh (carbon heavy), **b** 400 $gCO_{2e}$/kWh (moderately polluting), **c** 50 $gCO_{2e}$/kWh (clean/renewable)

# Decarbonization Scenarios

# 8

As discussed in previous chapters, numerous technological solutions are available for fulfilling humanity's energy needs with least environmental impact. Although currently available technologies are capable of providing the needed thermal energy at different energy efficiencies, the true environmental benefit is contingent upon primary energy's carbon intensity. Given this dependency, the following sections highlight the efficacy of different technologies in lowering the carbon footprint depending upon decarbonization of the primary energy supply.

## 8.1 Building Heating Technologies

Table 8.1 compares all heating technologies as a function of combined thermal and electrical carbon footprint while supplied with grid electricity at different carbon intensities and utilizing natural gas where applicable. The second row lists the electrical grid's carbon intensity while each column shows corresponding combined carbon footprint. The data can be interpreted visually by considering the filling within the circle icon or the color of the data point for the case being analyzed. Green color box represents lower carbon footprint while the yellow/red colors mean higher carbon footprint. As can be seen, in a grid decarbonization scenario, as the carbon intensity decreases, the heat pump-based solutions become favorable while the CHP based solutions are more impactful in regions with high electric grid carbon intensities. Within a single column representing a particular electric grid carbon footprint, the shift in combined carbon footprint can be noticed.

Similarly, Table 8.2 compares the combined carbon footprint possible with different heating technologies either supplied with electrical grid at a carbon intensity of 400

**Table 8.1** Electric grid decarbonization scenario—combined electrical and thermal energy carbon footprint in a building consuming 100 kWh/day thermal energy and 30 kWh/day electrical energy. Fuel carbon intensity of 180 gCO$_{2e}$/kWh

| Heating Technology | Electric Grid Carbon Intensity, gCO2e/kWh | | | | | |
|---|---|---|---|---|---|---|
| | 900 | 700 | 500 | 300 | 100 | 10 |
| Conv ER - 100% | 1827 | 1421 | 1015 | 609 | 203 | 20 |
| Conv EHP - 300% | 1227 | 954 | 682 | 409 | 136 | 14 |
| DFHP (60/40) - 220% | 1183 | 937 | 691 | 445 | 199 | 88 |
| SFFHP - 400% | 1152 | 896 | 640 | 384 | 128 | 13 |
| Sorption HP - 150% | 1047 | 841 | 635 | 429 | 223 | 130 |
| Conv FFD - 95% | 1116 | 910 | 704 | 498 | 292 | 200 |
| SPD - 95% | 1089 | 889 | 689 | 489 | 289 | 199 |
| CHP, no TES | 73 | 107 | 141 | 175 | 209 | 224 |
| CHP with sorp HP | 123 | 129 | 135 | 141 | 147 | 150 |
| CHP with TES | 115 | 138 | 160 | 183 | 205 | 215 |
| Blackstart HP, 30% el eff | 998 | 798 | 598 | 398 | 198 | 108 |
| Blackstart HP, 50% el eff | 973 | 773 | 573 | 373 | 173 | 83 |

gCO$_{2e}$/kWh and/or fuel with carbon intensities in the range of 300 gCO$_{2e}$/kWh (e.g., oil) to 10 gCO$_{2e}$/kWh (e.g., biofuels, hydrogen etc.), emulating the fuel decarbonization scenario. As shown, CHP technologies show the best performance, given the moderate carbon emissions from the electrical grid. Onsite cogeneration of electricity and thermal energy is beneficial in utilizing the total energy content of the primary energy efficiently. Amongst non-CHP technologies, sorption and blackstart heat pumps are better decarbonization options if the fuel's carbon intensity is below 100 gCO$_{2e}$/kWh.

Tables 8.3 and 8.4 show the carbon footprint associated with dual fuel systems operated at different ratios of primary energy supply (electrical grid vs. fuel supply) in a building consuming 100 kWh of thermal energy daily. Table 8.3 represents the electrical grid decarbonization scenario where the fuel is assumed as natural gas at a carbon intensity of 180 gCO$_{2e}$/kWh. As expected, the ratio shifts towards the primary energy source with lower carbon intensity. In the electrical grid decarbonization scenario, the threshold grid

**Table 8.2** Fuel grid decarbonization scenario—combined electrical and thermal energy carbon footprint in a building consuming 100 kWh/day thermal energy and 30 kWh/day electrical energy. Electric grid carbon intensity of 400 gCO$_{2e}$/kWh

| Heating Technology | Carbon Intensity of Fuel, gCO2e/kWh | | | | |
|---|---|---|---|---|---|
| | 300 | 200 | 100 | 50 | 10 |
| Conv ER - 100% | 812 | 812 | 812 | 812 | 812 |
| Conv EHP - 300% | 545 | 545 | 545 | 545 | 545 |
| DFHP (60/40) - 220% | 618 | 576 | 534 | 513 | 496 |
| SFFHP - 400% | 512 | 512 | 512 | 512 | 512 |
| Sorption HP - 150% | 612 | 545 | 479 | 445 | 419 |
| Conv FFD - 95% | 728 | 623 | 517 | 465 | 423 |
| SPD - 95% | 716 | 611 | 505 | 453 | 411 |
| CHP, no TES | 308 | 183 | 57 | -5 | -55 |
| CHP with sorp HP | 238 | 155 | 71 | 30 | -4 |
| CHP with TES | 316 | 195 | 75 | 15 | -33 |
| Blackstart HP, 30% el eff | 564 | 509 | 455 | 427 | 405 |
| Blackstart HP, 50% el eff | 522 | 482 | 441 | 420 | 404 |

**Table 8.3** Electric grid decarbonization scenario—thermal energy carbon footprint of dual fuel system operated at different primary energy operational ratios. Fuel carbon intensity of 180 gCO$_{2e}$/kWh

| DFHP (COP3), electric/fuel | Electric Grid Carbon Intensity, gCO2e/kWh | | | | |
|---|---|---|---|---|---|
| | 900 | 700 | 500 | 300 | 10 |
| 50/50 | 272 | 232 | 193 | 154 | 97 |
| 60/40 | 283 | 237 | 191 | 145 | 78 |
| 70/30 | 294 | 241 | 189 | 136 | 60 |
| 80/20 | 305 | 246 | 186 | 127 | 41 |
| 90/10 | 316 | 250 | 184 | 118 | 22 |

**Table 8.4** Fuel grid decarbonization scenario—thermal energy carbon footprint of dual fuel system operated at different primary energy operational ratios. Electric grid carbon intensity of 400 gCO$_{2e}$/kWh

| DFHP (COP3), electric/fuel | Carbon Intensity of Fuel, gCO2e/kWh | | | |
|---|---|---|---|---|
| | 300 | 200 | 100 | 10 |
| 50/50 | 220 | 167 | 115 | 67 |
| 60/40 | 198 | 156 | 114 | 76 |
| 70/30 | 177 | 145 | 114 | 85 |
| 80/20 | 155 | 134 | 113 | 94 |
| 90/10 | 134 | 123 | 113 | 103 |

carbon intensity is 500 gCO$_{2e}$/kWh where the primary energy ratio does not impact the overall carbon footprint. Similarly, in the fuel decarbonization scenario, the threshold carbon intensity of the fuel is 100 gCO$_{2e}$/kWh, where the primary energy ratio does not influence the overall carbon footprint.

## 8.2 Building Cooling Technologies

Tables 8.5 and 8.6 show the combined electrical and cooling energy associated carbon footprint in a building consuming 30 kWh per day and 100 kWh, respectively. Under electrical grid decarbonization scenario (Table 8.5), CHP based configurations show better carbon reduction potential compared to all other technologies at electrical grid carbon intensity above 300 gCO$_{2e}$/kWh. In fuel decarbonization scenario (Table 8.6), CHP integrated with sorption AC is the ideal solution under all fuel carbon intensity values even in regions with moderate electrical grid carbon intensities of 400 gCO$_{2e}$/kWh.

**Table 8.5** Electric grid decarbonization scenario—combined electrical and cooling energy carbon footprint. Fuel carbon intensity of 180 g$CO_{2e}$/kWh

| Cooling Technology | Electric grid carbon intensity, gCO2e/kWh | | | | | |
|---|---|---|---|---|---|---|
| | 900 | 700 | 500 | 300 | 100 | 10 |
| Conv AC - 300% | 1227 | 954 | 682 | 409 | 136 | 14 |
| Sorption AC - 70% | 1184 | 978 | 772 | 566 | 360 | 267 |
| CHP sorption AC 30% EL eff | 9 | 80 | 151 | 223 | 294 | 326 |
| BSAC, 30% El eff. | 1100 | 900 | 700 | 500 | 300 | 210 |
| BSAC 50% El eff | 1020 | 820 | 620 | 420 | 220 | 130 |
| CHP AC, 2 KW 30% eff | 341 | 313 | 285 | 258 | 230 | 218 |

**Table 8.6** Fuel grid decarbonization scenario—combined electrical and cooling energy carbon footprint. Electric grid carbon intensity of 400 g$CO_{2e}$/kWh

| Cooling Technology | Fuel grid carbon intensity, gCO2e/kWh | | | |
|---|---|---|---|---|
| | 300 | 200 | 100 | 10 |
| Conv AC - 300% | 545 | 545 | 545 | 545 |
| Sorption AC - 70% | 841 | 698 | 555 | 426 |
| CHP sorption AC 30% EL eff | 407 | 224 | 40 | -124 |
| BSAC, 30% El eff. | 733 | 622 | 511 | 411 |
| BSAC 50% El eff | 600 | 533 | 467 | 407 |
| CHP AC, 2 KW 30% eff | 416 | 296 | 175 | 67 |

## 8.3 Industrial Heating Technologies

Table 8.7 shows the total carbon footprint associated with both electrical and thermal energy consumption in an industrial complex located in regions with different electric grid carbon intensities while using electricity and/or natural gas as the primary energy sources. The table shows electrical grid decarbonization scenario where it compares the combined carbon intensity. It has to be noted that some of the systems are suitable for only low temperatures (e.g., heat pump systems). As can be seen, in the electrical grid decarbonization scenario, CHP configurations remain highly favorable until the grid carbon intensity falls below 100 g$CO_{2e}$/kWh. Similarly, CHP configurations offer lowest carbon intensity at all fuel carbon intensities even in moderate electrical grid carbon intensities of 400 g$CO_{2e}$/kWh (Table 8.8).

## 8.3 Industrial Heating Technologies

**Table 8.7** Electric grid decarbonization scenario—combined electrical and thermal energy carbon footprint. Fuel carbon intensity of 180 gCO$_{2e}$/kWh

| Heating Technology | Electric Grid Carbon Intensity, gCO2e/kWh | | | | | |
|---|---|---|---|---|---|---|
| | 900 | 700 | 500 | 300 | 100 | 10 |
| Conv ER - 100% | 1827 | 1421 | 1015 | 609 | 203 | 20 |
| Conv EHP - 300% | 1227 | 954 | 682 | 409 | 136 | 14 |
| DFHP (60/40) - 220% | 1183 | 937 | 691 | 445 | 199 | 88 |
| SFFHP - 400% | 1152 | 896 | 640 | 384 | 128 | 13 |
| Sorption HP - 150% | 1047 | 841 | 635 | 429 | 223 | 130 |
| Conv FFD - 95% | 1116 | 910 | 704 | 498 | 292 | 200 |
| CHP with TES | 207 | 208 | 208 | 208 | 209 | 209 |

**Table 8.8** Fuel grid decarbonization scenario—combined electrical and thermal energy carbon footprint. Electric grid carbon intensity of 400 gCO$_{2e}$/kWh

| Heating Technology | Carbon Intensity of Fuel, gCO2e/kWh | | | | | |
|---|---|---|---|---|---|---|
| | 300 | 200 | 150 | 100 | 50 | 10 |
| Conv ER - 100% | 812 | 812 | 812 | 812 | 812 | 812 |
| Conv EHP - 300% | 545 | 545 | 545 | 545 | 545 | 545 |
| DFHP (60/40) - 220% | 618 | 576 | 555 | 534 | 513 | 496 |
| SFFHP - 400% | 512 | 512 | 512 | 512 | 512 | 512 |
| Sorption HP - 150% | 612 | 545 | 512 | 479 | 445 | 419 |
| Conv FFD - 95% | 728 | 623 | 570 | 517 | 465 | 423 |
| CHP with TES | 348 | 231 | 173 | 115 | 57 | 11 |

# Concluding Remarks

Global energy demand growth and climate urgency are driving the need for efficient utilization of available primary energy with least environmental impact. Buildings and industries together consume ~75% of primary energy, globally. Decarbonization of these sectors requires careful consideration towards the carbon intensity of the primary energy. This book provides a historical context on the emergence of energy sources and consumption sectors. The role of different primary energy sources in decarbonizing individual sectors utilizing conventional and emerging technologies is discussed in detail. The global ambition to achieve a net-zero economy in the next few decades is expected to make progress at different paces, dictated by individual country's economic development plan, access to capital, and availability of desired primary energy resources. The methodology employed here can be easily customized for different regions of the world to quickly assess the decarbonization potential of individual power and thermal technologies with different primary energy resources. The impact of electrical grid and fuel grid decarbonization scenario on the overall carbon intensity of delivered energy is provided. Adoption and deployment decisions can be guided by projected growth in energy demand, marginal emission factors, resiliency considerations, time-of-use charges along with capital costs, retrofit challenges, and primary energy flexibility.

# Nomenclature

| Parameter | Description | Units |
|---|---|---|
| $Load_{TH}$ | Building thermal load | kWh/day |
| $\eta_{Thermal\ technology}$ | Energy efficiency of the thermal technology being analyzed—in both cooling (if applicable) and heating modes of operation | % |
| $CO2_{grid}$ | Equivalent carbon dioxide emission factor of the power grid | g/kWh |
| $CO2_{fuel}$ | Equivalent carbon dioxide emission factor of fuel | g/kWh |
| $\kappa$ | Balance of plant parasitic power for the heating technology (e.g., air handler) | % of thermal load |
| $\eta_{chp,EL}$ | Electrical efficiency of the combined heat and power system | % of higher heating value of the fuel |
| $Load_{EL}$ | Building electrical load | kWh/day |
| $\gamma_{heat\ recovery}$ | Useful heat recovery from CHP based configurations | % of total useful heat available |
| $t_{operation}$ | Operational hours of CHP per day | Hours/day |
| $kW_{chp\ TES}$ | Power rating of the CHP in thermal energy storage configuration | kW |
| $Load_{TH,true}$ | Actual primary thermal energy needed | kWh/day |
| $x_{el}$ | Fraction of electrical driven operational hours in a dual fuel configuration | % |
| $x_{fuel}$ | Fraction of fuel driven operational hours in a dual fuel configuration | % |
| $CO2_{TH}$ | Total equivalent carbon dioxide emissions of the thermal system including balance of plant | g/kWh |

(continued)

(continued)

| Parameter | Description | Units |
|---|---|---|
| $CO2_{TH+EL}$ | Total equivalent carbon dioxide emissions of the combined thermal and electrical load of the building | g/kWh |
| $CO2_{EL}$ | Total equivalent carbon dioxide emissions associated with building electrical load | g/kWh |
| $CHP_{TH\,rating}$ | Thermal power rating needed from the CHP (without TES option) | kW |
| $CHP_{EL,gen}$ | Electrical energy generated by the CHP in the configuration without TES | kWh/day |
| $\beta_{EL,exp}$ | Electrical energy exported to the power grid | kWh/day |
| $\upsilon_{EL,exp,TES}$ | Electrical energy exported to the power grid in CHP configurations with TES | kWh/day |
| $CO2_{chp}$ | Total equivalent carbon dioxide emissions associated with building thermal and electrical load (for configurations without TES) | g/kWh |
| $CHP_{TH,TES}$ | Thermal energy generated from CHP in configurations integrated with TES | kWh/day |
| $CO2_{chp,TES}$ | Total equivalent carbon dioxide emissions associated with building thermal and electrical load (for configurations with TES) | g/kWh |
| $\alpha$ | Fraction of electric energy dissipated as useful heat | % |
| $f_{PC}$ | Fuel supply to power cycle | % |
| $\eta_{PC}$ | Power cycle conversion efficiency for self-powered devices | % |
| $COP_h$ | Heat pump's heating COP | Non-dimensional |
| $\eta_{SPD}$ | Calculated energy efficiency of the self-powered device | % |
| $\eta_{SPD-HP}$ | Calculated energy efficiency of the self-powered heat pump device | % |
| $kW_{EL,BSHP}$ | Electrical power rating of the prime mover for black start heat pump | kW |
| $Load_{TH,BSHP}$ | Thermal energy needed by the black start heat pump after heat recovery from the prime mover | kWh/day |

(continued)

# Nomenclature

(continued)

| Parameter | Description | Units |
|---|---|---|
| $Load_{TH,BSHP,true}$ | Actual electrical energy needed by the black start heat pump after heat recovery from the prime mover | kWh/day |
| $kW_{TH,BSHP}$ | Thermal power from the PM of black start heat pump | kW |
| $CHP_{EL,BSHP}$ | Electrical energy generated by the PM of black start heat pump | kWh/day |
| $CO2_{TH,BSHP}$ | Total equivalent carbon dioxide emissions associated with building thermal load provided by black start heat pump configuration | g/kWh |
| $CO2_{chp\,sorption AC}$ | Total equivalent carbon dioxide emissions associated with building thermal and electrical load installed with CHP based sorption AC configurations | g/kWh |
| $CHP_{EL,ConvAC}$ | Electrical power rating of the CHP integrated with conventional electric AC | kW |
| $COP_c$ | Electric AC's cooling efficiency | Non-dimensional |
| $CO2_{CHP\,conv\,AC}$ | Total equivalent carbon dioxide emissions associated with building thermal and electrical load installed with CHP based conventional AC configurations | g/kWh |
| $kW_{EL,BSAC}$ | Electrical power rating of the prime mover for black start conventional AC | kW |
| $CO2_{TH,BSAC}$ | Total equivalent carbon dioxide emissions associated with building thermal and electrical load installed with CHP based black-start AC configurations | g/kWh |

# References

1. *9HA Gas Turbine*. (n.d.). Retrieved April 5, 2021, from https://www.ge.com/gas-power/products/gas-turbines/9ha.
2. Smil, V. (2003). *Energy at the crossroads*. MIT Press.
3. U.S. Energy Information Admnistration. (n.d.-a). *Table 8.2. Average Tested Heat Rates by Prime Mover and Energy Source, 2009–2019*. https://www.eia.gov/electricity/annual/html/epa_08_02.html.
4. U.S. Energy Information Admnistration. (n.d.-b). *U.S. Shale Production*. https://www.eia.gov/dnav/ng/hist/res_epg0_r5302_nus_bcfa.htm.
5. U.S. Energy Information Admnistration. (2020). *Net Generation by Energy Source: Total (All Sectors), 2009–2019*. https://www.eia.gov/electricity/annual/.
6. *U.S. Greenhouse Gas Emissions from Electricity Generation*. (n.d.). https://cfpub.epa.gov/ghgdata/inventoryexplorer/#electricitygeneration/entiresector/allgas/category/all.
7. Vandervort, C. (2018). Advancements in H class gas turbines and combined cycle power plants. *Proceedings of the ASME Turbo Expo, 3*, 1–10. https://doi.org/10.1115/GT201876911.
8. Yi, Y. (2003). Simulation of a 220 kW Hybrid SOFC Gas Turbine System and Data Comparison. *ECS Proceedings Volumes, 2003–07*(1), 1442–1454. https://doi.org/10.1149/200307.1442pv.
9. International Energy Agency (2023). "Energy Technology Perspectives."
10. Dudley, B. (2023). Statistical Review of World Energy. Centre for energy economics research and policy. British Petroleum, Available via https://www.bp.com/en/global/corporate/energy-economics/statistical-review-of-world-energy.html.
11. International Energy Agency (2022). "World energy outlook 2022." Paris, France.
12. Lenzen, Manfred. "Life cycle energy and greenhouse gas emissions of nuclear energy: A review." Energy conversion and management 49.8 (2008): 2178–2199.
13. Amponsah, Nana Yaw, et al. "Greenhouse gas emissions from renewable energy sources: A review of lifecycle considerations." *Renewable and Sustainable Energy Reviews* 39 (2014): 461–475.
14. Hannah Ritchie, Max Roser and Pablo Rosado (2022) – "Energy". Published online at OurWorldInData.org. Retrieved from: https://ourworldindata.org/energy.
15. Gluesenkamp, K.R., et al., *Self-powered Heating: Efficiency Analysis*. 2021.
16. Cheekatamarla, P., Kowalski, S., Abu-Heiba, A., LaClair, T., & Gluesenkamp, K. , *Modeling and Analysis of a Thermophotovoltaic Integrated Self-Powered Furnace*. Energies, 2022. **15**(19): p. 7090.
17. Abu-Heiba, A., et al., *Analysis of power conversion technology options for a self-powered furnace*. Applied Thermal Engineering, 2021. **188**: p. 116627.

18. LaPotin, A., et al., *Thermophotovoltaic efficiency of 40%.* Nature, 2022. **604**(7905): p. 287–291.
19. Cheekatamarla, P., *Role of On-Site Generation in Carbon Emissions and Utility Bill Savings under Different Electric Grid Scenarios.* Energies, 2022. **15**(10): p. 3477.
20. Cheekatamarla, P., *Performance analysis of hybrid power configurations: Impact on primary energy intensity, carbon dioxide emissions, and life cycle costs.* International Journal of Hydrogen Energy, 2020. **45**(58): p. 34089–34098.
21. Li, Z., *Hybrid Heat Pump Controls: Conventional Dual Fuel versus Seamlessly Fuel Flexible Heat Pump.* 2022.
22. Li, Z., et al., *Seamlessly Fuel Flexible Heat Pump with Optimal Model-based Control Strategies to Reduce Peak Demand, Utility Cost and CO2 Emission.* 2022, Oak Ridge National Lab.(ORNL), Oak Ridge, TN (United States).
23. International Energy Agency, I., *Global Energy Review: CO2 Emissions in 2021.* 2022.
24. Carbon Leadership Forum, C.C. *Electricity Grid Carbon Intensity - Canada, US, Global.* 2021; Available from: https://community.carbonleadershipforum.org/t/electricity-grid-carbon-intensity-canada-us-global/2440.
25. Volker-quaschning. *Specific Carbon Dioxide Emissions of Various Fuels.* 2022; Available from: https://www.volker-quaschning.de/datserv/CO2-spez/index_e.php.
26. International Energy Agency, I., *The Future of Cooling.* 2022.
27. Administration, U.E.I. *Annual Energy Outlook, 2021.* 2021 [cited 2021 July 3rd]; Available from: https://www.eia.gov/outlooks/aeo/tables_ref.php.
28. National Academies of Sciences, E., and Medicine, *Accelerating Decarbonization in the United States Energy Sector.* 2021.
29. Energy, U.S.D.o. *Quadrennial Technology Review 2015, Technology Assessment 6I: Process Heating Systems.* 2015; Available from: https://www.energy.gov/sites/prod/files/2016/06/f32/QTR2015-6I-Process-Heating.pdf.
30. Office, U.S.D.o.E.A.M. *Manufacturing Energy and Carbon Footprints (2018 MECS).* 2018; Available from: https://www.energy.gov/eere/amo/manufacturing-energy-and-carbon-footprints-2018-mecs.

GPSR Compliance

The European Union's (EU) General Product Safety Regulation (GPSR) is a set of rules that requires consumer products to be safe and our obligations to ensure this.

If you have any concerns about our products, you can contact us on ProductSafety@springernature.com

In case Publisher is established outside the EU, the EU authorized representative is:

Springer Nature Customer Service Center GmbH
Europaplatz 3
69115 Heidelberg, Germany

**Batch number: 09151338**

Printed by Printforce, the Netherlands